History and
Philosophy
of Biology

History and Philosophy of Biology

Robert Kretsinger

University of Virginia, USA

World Scientific

NEW JERSEY • LONDON • SINGAPORE • BEIJING • SHANGHAI • HONG KONG • TAIPEI • CHENNAI

Published by

World Scientific Publishing Co. Pte. Ltd.
5 Toh Tuck Link, Singapore 596224
USA office: 27 Warren Street, Suite 401-402, Hackensack, NJ 07601
UK office: 57 Shelton Street, Covent Garden, London WC2H 9HE

Library of Congress Cataloging-in-Publication Data
Kretsinger, Robert H.
  History and philosophy of biology / Robert Kretsinger, University of Virginia, USA.
    pages cm
  ISBN 978-9814635035 (hardcover : alk. paper) -- ISBN 9814635030 (hardcover : alk. paper) --
ISBN 978-9814635042 (pbk. : alk. paper) -- ISBN 9814635049 (pbk. : alk. paper)
  1. Biology--Philosophy. 2. Biology--History. I. Title.
  QH331.K727 2015
  570.1--dc23

                              2014050204

British Library Cataloguing-in-Publication Data
A catalogue record for this book is available from the British Library.

Acknowledgments

Ms Sine Harris, Glasgow, provided the figure for "genes or environment" (Chapter 46). Carter Ransom suggested many changes to make the discussions more accessible to the general reader. Many scholars, only a few of whom are explicitly cited, did the fundamental research that provided the foundation for HPB; thank you.

Contents

Section A

History and Philosophy: Overview

How has one thought about science in times past and in various cultures? What are the more meaningful and rewarding ways to think about science today? Most definitions or characterizations of science fall into two categories:

The first says that science is the study of the natural world. If one includes applications of science, then immediately one is dealing with an un-natural world — one that reflects human activities, especially engineering and medicine. More difficult is the concept of the natural world. Surely investigating the blood circulation of a mouse is science, as is the study of its mating behavior. One then asks whether studying the circulatory system of a human is science. How about the study of his mating behavior? Of his art?

The second definition emphasizes procedure or technique. Does science involve the formulation of a theory and its subsequent testing? If so, how does one go about testing a theory in a historical science such as geology or biology? What distinguishes generalizations from hypotheses, from theories, from laws? Is there a scientific method(s); and if so, should it be applied to the study of human activities such as economics, governance, and art?

The following chapters address these questions: What is science? How have various cultures thought about science? What are contemporary perspectives on science and how have they evolved in recent centuries?

A1. Pre-Hellenic Science
A2. Hellenic Science
A3. Chinese Science
A4. Islamic Science
A5. Early Christianity
A6. Inductive Logic, "Works," and Francis Bacon
A7. Deductive Logic, Maths, and René Descartes
A8. The Scientific Revolution
A9. The Church and Science
A10. Falsifiability: Karl Popper
A11. Paradigm: Thomas Kuhn
A12. Two Cultures: C.P. Snow
A13. Emergence

A1

Pre-Hellenic Science

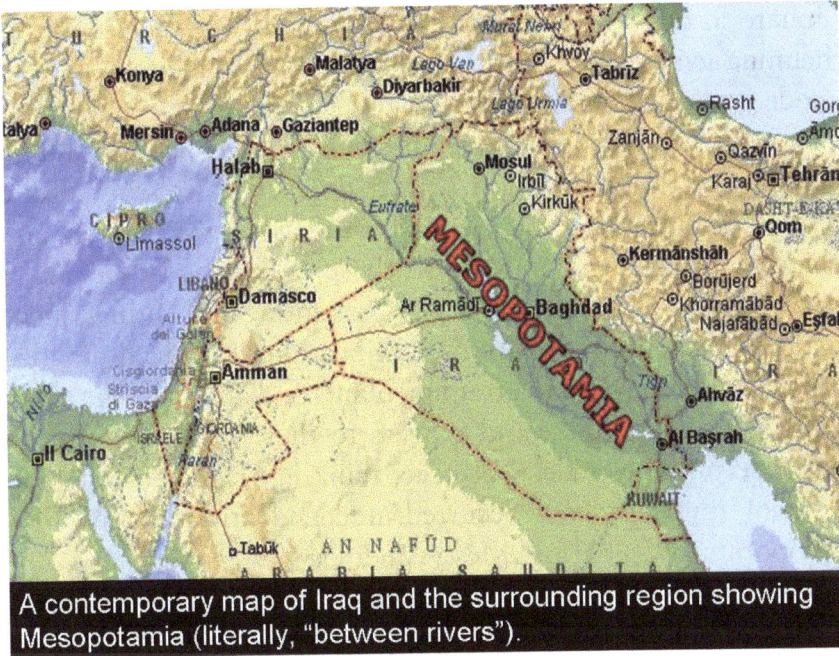

A contemporary map of Iraq and the surrounding region showing Mesopotamia (literally, "between rivers").

As will be discussed in the next chapter, most historians, especially those in the West, appropriately turn to Greece to understand the origins of science, as well as many other intellectual endeavors. However, there are strong arguments for including engineering, informatics, medicine, and agriculture in the definition of science. If so, we should consider our first tools, garments, dwellings, and herbal remedies.

3

It requires reasonable talent to flake a stone and attach it to a handle to make an axe. It is hardly trivial to cure a hide and sew pieces together to make a cape or shoes or to bind branches and leaves to make a sturdy shelter. Chimpanzees use digging sticks to access termite nests. Other mammals and birds use tools and make complex nests and burrows. Macaws intuitively ingest clay, apparently to absorb toxins in some seeds that they eat. These skills are to some extent transmitted by imitation, but most are inherent. It seems reasonable to infer that some ability to do simple science is innate in our own species, as well as in others. There is overwhelming evidence that other complex behaviors are genetically encoded — so much for tabla rasa.

One can only speculate about the development of human language and the urges to do art and to do science. These intellectual abilities are probably inter-related. In any case, several fundamental achievements occurred independently several times in human evolution. These include the concepts of counting and adding, identification of stellar constellations, codification of medical practices, as well as the naming and grouping plants and animals. The assignment of names, stories, and powers to these constellations, animals, and practices reflects abstract thinking. These basic achievements of applied science occurred independently in Egypt, India, Mesopotamia, China, and Meso-America.

The earliest astronomer for whom we have records is Thales of Miletus (~600 B.C.). We are left to marvel as to why the Greeks extended these speculations and analyses to new levels of sophistication and abstraction. Did their achievements reflect unique environmental circumstances or the convergence of yet to be identified historical currents? Or were the circumstances that led to the appreciation of questioning purely stochastic? This is hardly a popular interpretation. However, the antecedents, if any, of Greek philosophy have yet to be established.

This chapter summarizes some of the early achievements of the Egyptians, Indians, Assyrians, and Babylonians — peoples of the Bronze Age in the Middle East. Subsequent chapters survey Chinese and Islamic science.

One refers to Egypt, after the unification of the upper and lower kingdoms about 3000 B.C., without exploring the subtleties of different dynasties. Lunar and solar calendars were merged; their calendar consisted of 12 months each of 30 days plus five special days committed to religious holidays. Sundials gave a precise definition of the solstices.

Much of their knowledge of anatomy came from mummifiers, who inserted a long hook through a nostril, broke the thin ethmoid and removed the brain. They removed viscera through an incision in the left groin. These procedures seemed not to have laid the foundation for further exploration of anatomy. Why were they not more curious?

The Ebers Papyrus (~1550 B.C.) listed some 877 "prescriptions" and noted a "... tumor against the god Xenus ... do nothing there against." Homer (~700 B.C.) in the Odyssey noted that "... the Egyptians were skilled in medicine more than any other art." Herodotus (484–425) visited Egypt ~440 B.C. and wrote of their advanced medical practices. Pliny the Elder (23–79) praised their medicine. However, they failed to distinguish arteries from veins or nerves from tendons. The heart was assigned spirit and thought. Hippocrates, Herophilos, Erasistratus, and Galen studied at the temple to Amenhotep III across the Nile from Luxor. Peseshet (~2400 B.C.), mother of Akhethotep, was the first female doctor on record.

The Egyptians made potions or amulets with animal or plant parts that resembled afflicted regions — "*simila similibus*" (similar with similar), a concept not unknown to modern homeopathy. They distinguished phylactic, protection against demons, from theophoric procedures that invoked the help of a deity.

Herodotus in his *Histories* noted that circumcision was the norm and that the Egyptian military brought back uncircumcised phalli of Libyans as souvenirs. How might one evaluate the effectiveness, physiologically or psychologically, of their medicine? Or is effectiveness the right question?

The Edwin Smith papyrus (~1550 B.C.), Hearst papyrus (~1450 B.C.), and Berlin papyrus (~1200 B.C.) noted a range of foods and a sophisticated agriculture that reflected the importation of plants and animals from thousands of kilometers — "... milk, three kinds of beer, five kinds of wine, ten loaves, four of bread, ten of cakes, four meats, different cuts, joints, roast, spleen, limb, breast, quail, goose, pigeon, figs, ten other fruits, three kinds of corn, barley, spelt, five kinds of oil, and fresh plants ..."

After the annual flood of the Nile, fields had to be re-surveyed; they made right angles using 3, 4, 5 triangles. We still marvel at their feats of civil engineering — pyramids, obelisks, the fabled light tower at Alexandria, and complex irrigation systems. Their ships could sail 90° to the wind. They made jars from molten glass. They made quality papyrus (paper) from reeds and developed hieroglyphs with phonetic symbols. Egyptians mastered a lot of engineering and agriculture. Many of these practices seemed to have remained unchanged from ~3400 B.C. until the Persian invasion of 525 B.C.

The people of the Indus valley, prior to 500 B.C., developed a calendar of 12 months, 30 days per month, with an intercalary month as needed about every sixth year. Their math incorporated zero and a base 10 number system; it included sine and cosine tabulations. Their metallurgy produced large cast iron pillars. They made stainless steel (wootz, with particles of Fe_3C) sword blades, later called Damascus steel in the West. Mercury and sulfur were used in metallurgy and as medicines. Several medical texts or vedas were compiled.

Just as Egypt developed beside the Nile and India on the banks of the Indus, so Mesopotamia developed between the Tigris and the Euphrates rivers in present day Iraq and southeast Turkey. The succession of peoples, languages, and rulers is complex. The important point is that by ~500 B.C. they had made significant advances. They developed a base 60 numeral system — hence our 60 minute hour, 24 hour day, and 360° circle. Al-Batani reckoned the precession of the earth's axis of rotation to be 54.5 arc-seconds per year; this compares well to the current value of 49.8 (see Chapter B3).

They were among the first to make quality bronze, cloth woven of wool and flax, and complex irrigation systems. Esagil-kin-apli of Borsippa wrote one of several *Diagnostic Handbooks* about 1050 B.C.

Egypt, India, Mesopotamia, and China (to be discussed in Chapter A3) all reached reasonable levels of sophistication with limited inter-communication. The details of their sciences varied. However, one can see that given a bit of political stability and economic self-sufficiency the pursuit of science and its applications seems inherent. These advances occurred before Thales (~624–~546 B.C., see Chapter A2) and a millennium of Greek leadership in inquiry. Islamic science (Chapter A4) built on the heritage of Egypt and Mesopotamia. One might then ponder why the Greeks tolerated all sorts of contentious speculations and why only in post-renaissance Europe did science proceed to higher levels of abstraction and sophistication.

A2

Hellenic Science

The School of Athens (1509).

The term "Hellenic" refers to both Greek language and Greek culture. Their civilization extended from Macedonia to southern Italy including Sicily, to Egypt, and to cities near the Mediterranean coast of present Turkey and Syria. Significant insights and innovations were made in mathematics, astronomy, physics, anatomy, and botany from 600 B.C. to 400 A.D. The empire of Alexander (356–323 B.C.) was fragmented soon after his death. It was not a unified kingdom with a single ruler or council and therein may have laid its intellectual vitality.

The Greeks were not the first to address abstract philosophies. However, what set them apart was their tolerance of, even pleasure in, disputation. This intellectual freedom was more limited in authoritarian or monarchical regimes. They posed questions, still relevant today, about the nature of knowledge.

This chapter provides a brief summary of these achievements. The impact of Greek mathematics, astronomy, and architecture on the Roman Empire, on the Islamic world, then on Western Europe are irrefutable. More problematic are their views on motion, the void, and atoms. One may question whether the "Great Chain of Being" of Aristotle or the anatomy of Galen advanced or hindered understanding of biology. This review of just their natural philosophy, or science, does not capture the full impact of their thinking.

Thales (~624–~546 B.C.) understood similar and right triangles; he calculated the height of a pyramid from the length of
its shadow. Bertrand Russell opined that "Western Philosophy
begins with Thales." Pythagoras (~575–~495 B.C.) is "...the
father of numbers." He argued "...number is the ruler of forms
and ideas and the cause of gods and demons." He gave the
first proof of $a^2 + b^2 = c^2$; he realized that $2^{0.5}$ is irrational
(Chapter B2). He analyzed vibrating strings and deduced that tones
of a musical scale could be described as frequencies related as the
ratio of whole numbers (Chapter D5). Euclid (~300 B.C.) is "...the
father of geometry." His *Elements* consists of 13 books and 36
propositions; it is the "...most important book of mathematics
ever written." He also helped lay the foundations of number theory. Archimedes (~287–~212 B.C.) brought mathematical analysis to engineering. He analyzed the block and tackle as well as
levers. "Give me a place to stand on, and I will move the Earth."
He calculated the value of $\varLambda$ by inner and outer polygons of 96 sides
to be between $3 + 1/7$ (~3.1429) and $3 + 10/71$ (3.1408). Hypatia
(~360–~415 A.D.) was the first female mathematician of record.
   Thales supposedly predicted a solar eclipse. Philolaus (~480–
~385 B.C.), as cited by Copernicus, "...knew that the Earth
revolves around a central fire." Plato (427–347 B.C.) wrote in *The
Republic*: "We shall approach astronomy, as we do geometry, by
way of problems, and ignore what's in the sky, if we intend to
get a real grasp of astronomy." Aristarchus (~287–~212 B.C.)
also argued a heliocentric model but could not detect the predicted
parallax of distant stars. Eratosthenes (~276–~195 B.C.) made a
map of the known (Mediterranean) world and developed a system of latitude and longitude. He calculated the circumference of
the earth based on the angle of elevation of the sun at noon on
the summer solstice as well as the tilt of the Earth's axis (23.4°,
Chapter B3). Hipparchus is regarded as the greatest astronomer

of antiquity; he completed the first comprehensive star catalog in the West. He developed spherical trigonometry and made accurate models of the motion of the sun and moon based on the concept of epicycles. He discovered the precession of the moon and estimated the eccentricity of the solar orbit. Proclus(412–485) made the last recorded astronomical observation of the Greeks in 475. It was a good millennium.

Thales argued that all matter is one, basically water. But how then could it exhibit so many properties? Anaximander (610–546 B.C.) adopted the concept of four elements (air, earth, water, and fire). Leucippus (~475 B.C.) explored the idea of atoms and empty spaces between them to permit motion. Parmenides (~515–~440 B.C.), in contrast, argued that the void is nothing; it offers no resistance, hence infinite speed and therefore movement is impossible. He also explored the duality of appearance and reality and concluded that truth cannot be known via sensory perception; only by pure reason, logos. All of this, long before Descartes (Chapter A7).

Democritus (~460–~370 B.C.) a student of Leucippus, elaborated on the nature of the atoma, "indivisible units," and argued that "...atoms and the void alone exist." However, he did not relate his atoms to air, earth, fire, and water. Aristotle (384–322 B.C.) attributed properties to air, wet and hot; to earth, dry and cold; to water, cold and wet; and to fire, hot and dry. He was aware of elements that we now know as sulfur (S), iron (Fe), copper (Cu), silver (Ag), tin (Sn), gold (Au), mercury (Hg), lead (Pb), and probably arsenic (As), antimony (Sb), bismuth (Bi), as well as numerous compounds: water, salt, acid, lye, alum, ochre, cinnabar, oil, pitch, steel, natron, wine, litharge, bronze, lime, vinegar. This seemingly inconsistent view — earth, air, fire, and water vs. elements — was not questioned.

Some anatomy and physiology can be inferred without experimentation. Hippocrates (~460–~370 B.C.), the "father of medicine," had no access to human dissections. He understood physiology in terms of the four humors — blood, black bile, yellow bile and phlegm — and rationalized diseases in terms of imbalances of these humors, or dyscrasia. He proposed standards of practice and distinguished between diagnosis, often to permit the family to plan, and treatment of which he had few. Aristotle (384–322 B.C.) distinguished aquatic mammals from fish. He noted stages of the development of the chick embryo, as well as of the mammal-like embryology of the hound shark, *Mustelus laevis*.

Herophilos (335–~280 B.C.) performed the first recorded dissections of humans, executed criminals, as cited by Galen. He distinguished motor from sensory nerves and assigned the site of intelligence to the brain. Erasistratus (304–~250 B.C.), a colleague of Herophilos in the school of anatomy in Alexandria, identified valves in the heart, recorded palpitations, and assigned its function as a pump. Galen (129–~208 A.D.) was a surgeon in the gladiator school of Pergamon; this provided him with "windows into the body." He also dissected various animals, including the macaque (barbary ape). His writings, with a few errors, became the reference point for copy or criticism by various Islamic and medieval anatomists. Herophilos and Erasistratus established a school of anatomy school in Alexandria where human dissections were permitted. They distinguished nerves from blood vessels and motor from sensory nerves.

Aristotle referred to observations of botany and zoology, as well as to experiments: "Salt water when it turns into vapor becomes sweet, and the vapor does not form salt water when it condenses again. This I know by experiment." He accepted various deities but distinguished logos from mythos. He suggested that plants have a vegetative soul and that animals have a sensitive soul. Humans are unique in having a spiritual soul. He sought a perfect representative of each species, the essence of typographic thinking, and the "Great Chain of Being" leading to humans at the pinnacle. One of the great challenges to biology of the scientific revolution was to refute many of these ideas and their overly simplistic interpretations. Theophrastus (370–~285 B.C.) was the guardian of Aristotle's children and succeeded him as leader of the peripatetic school where he argued against the "Great Chain of Being." He is credited as the first botanist to attempt some sort of classification — *Enquiry into Plants*, nine books, and *On the Causes of Plants*, six books. Pedanius Dioscorides (~40–~90 A.D.), a physician in Rome, wrote the first pharmacopeia, *De Materia Medica*, five volumes (Chapter C10).

In Raphael's *School of Athens*, Plato's hand points upward toward the heavens, Aristotle's down towards the Earth — idea and theory vs. observation and evidence. This image captures one of the great questions, still relevant today, addressed by the Greeks. Their achievements, from math through botany, comprise an impressive scientific legacy. More important, they posed abstract questions and sought general principles. Although the Greek worshipped many deities, they distinguished between mythos and logos and maintained that nature is ruled by laws (Chapters A5, A9, D3). Pythagorus argued that "... number is the ruler of forms and ideas and the cause of gods and demons." Many favored explanations that could be related to whole numbers or ideal solids. Were the stars fixed to the inner surface of a vast sphere? Could there exist anything beyond that, and how did this relate to infinity?

What of nature could be learned from observation? Would not experimentation "vex" the system and render it no longer natural? The attempts to make sense of the observed plants and animals begged one of the fundamental questions of biology: What is the appropriate or "natural" order of these organisms? Aristotle presented his "Great Chain of Being" as a hierarchy with insensate plants at the bottom, then soulless animals, topped by human beings. This led naturally enough to a ranking of humans, not for the first or the last time, with the home team inevitably at the top.

A3

China and Early Science

Zhang Qian travels to the West (Tang, 618–712).

The recorded history of China predates that of Greece and perhaps Egypt. China developed with minimal communication with the Mediterranean world. Many Chinese discoveries and applications of science predated the corresponding events in the West. The West had heard rumors about China long before Marco Polo traveled the silk routes. Francis Bacon referred to their great achievements — gunpowder, paper, printing, and the compass. Any simple enumeration losses detail and nuance; nonetheless, the overall impact is irrefutable. Between 500 and 1500 China was ahead of Islam and Europe in science.

This generalization suggests several inter-related questions. How or why did China make these inventions? Was their approach to science inherently different from that of the Greeks or of medieval Islam or Europe? These considerations lead to the Needham question: "Given the advances in science in China up to ~1600, why did science then stagnate?" This, in turn, begs the inverse Needham question: "Why did Europe enjoy such a fluorescence of creative energy during the renaissance and the subsequent scientific revolution?"

Joseph Needham (1900–1995), an embryologist in Cambridge, England, directed from 1942–1946 the Sino-British Science Co-operation Office in Chongqing, in central southern China well beyond the reach of the Japanese army (Winchester S. Bomb, book and compass: The fantastic story of the eccentric scientist who unlocked the mysteries of the Middle Kingdom, 2008). In addition to providing liaison and support to Chinese scientists, he traveled in western China and gathered information and artifacts related to the history of science. He subsequently founded the Needham Research Institute in Cambridge; it is still documenting the history of science in China. Many scholars have addressed the Needham question "Given the advances in science in China up to ~1600, why did science then stagnate?" Most interpretations include: "A general decline in the economy and vitality of the Ming (1368–~1650) and Qing Dynasties (1644–1912) was reflected in diminished support for and interest in science." "The science of the preceding millennium was focused on application as opposed to abstractions; hence, could not advance." This discussion, then introduces the themes of Chapters A4 and A5 "Why did science suddenly flourish in the world of Islam and in Europe following the renaissance?"

The record of early scientific advances and technical applications is impressive. To the dismay of scholars, overviews of Chinese science often present a list of nominal accomplishments without exploring the details of the device or procedure. Such listings, superficial as they may be, do support several generalizations:

China developed productive agriculture by 6000 B.C. — wheat in the North, rice in the South — and an ensemble of related Mandarin dialects by ~3000 B.C., seemingly without knowledge of parallel developments in Egypt and Mesopotamia. China realized advances in engineering and medicine well before their counterparts in the Middle East or Europe. Its industrial might was impressively demonstrated by Qin Shi Huang (259–210 B.C.), first emperor of united China. He commanded the production of 8,000 terra cotta soldiers and 700 horses, discovered in 1970 near his mausoleum. How would one compare this, as a fraction of gross domestic product, with the Pyramid of Giza?

A few traders from the Middle East and India plied the silk route(s) to China before ~1000 A.D. Others traded with the Spice Islands. The existence of the Far East was rumored in medieval Europe. The Mongols, from Genghis Khan (1162–1227) to his grandson, Kublai Khan (1215–1294), reigned over most of China for a century. The accounts of Marco Polo (1254–1324) were widely circulated upon his return in 1295 from a 24 year sojourn. By 1549, Portuguese and Spanish missionaries were established in Nagasaki, Japan, and by 1582 in Macao, China.

Zheng-Ho of the Ming Dynasty (1368–1644) sent seven expeditions from the South China Sea, across the Persian Gulf, to the east coast of Africa (1405–1433). Among other artifacts, they brought back a giraffe. After the seventh, he banned further exploration, "been there, done that," and explored no more (until the last few decades, making up for lost time). In contrast to the Portuguese and Spanish, he did not send missionaries.

Although medieval citation indices are not so thorough or easily accessed, Francis Bacon (1561–1626) referred to the advanced state of Chinese civilization and specifically cited their inventions of "bombs, books, and compasses." The earliest reference to firecrackers is in 290 A.D., to a fire lance in 950, and to bombs launched from trebuchets in 1161. These were supposedly based on a black powder — charcoal, sulfur, and potassium nitrate. The *Diamond Sutra*, recovered from one of the Mangao Caves (Caves of a Thousand Buddhas) in 868 was block printed on paper made from wood pulp; it is a Chinese translation of a Sanskrit praise of Buddha. The use of naturally occurring magnetite (Fe_3O_4) rocks as compasses in navigation was recorded in 1116.

The earliest pottery has been dated to 10,000 B.C., seemingly before agriculture was established. Potsherds have been found in both the Yangzi and Yellow river basins, evoking comparison with the Nile, Tigris, Euphrates, and Indus. One succumbs to tabulating just a few of their most impressive early inventions:

alcohol distillation	1500 B.C.
acupuncture	580
refraction of light	400
antimalarial, artemisinin	300
mercury distillation	300
diurnal rhythms and disease	200
circulation, arterial v. venous	200
goiter, treatment with iodine	100

The point of this discussion is not to compare each discovery with its counterpart in pre-Renaissance Europe. Most contemporary historians would agree with Bacon that overall the engineering in China was better than that in Europe. Further, it had been realized independently with little or no knowledge of the technical or of the philosophical achievements of the Mediterranean world.

These examples beg the fundamental question of whether and how society influences science and conversely how science affects society.

A4

Islamic Science

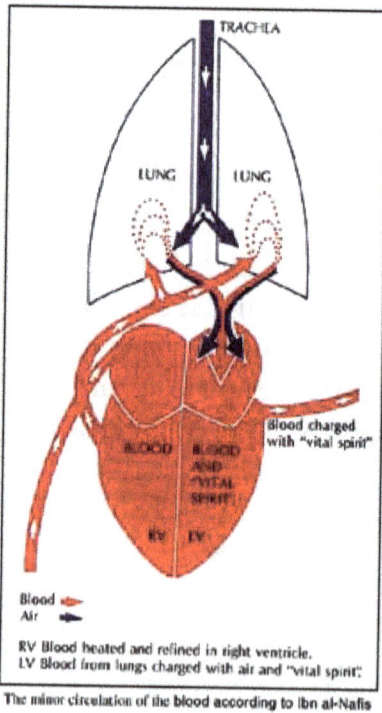

TRACHEA

LUNG   LUNG

Blood charged
with "vital spirit"

BLOOD   BLOOD
         AND
         "VITAL
         SPIRIT"

RV   LV

Blood
Air

RV Blood heated and refined in right ventricle.
LV Blood from lungs charged with air and "vital spirit".

The minor circulation of the blood according to Ibn al-Nafis

Circulatory system.

During the century following the death of the prophet Muhammad in 632, Islam spread across North Africa into Spain and throughout the Middle East into North India. The apogee of that expansion might be considered to be the conquest of Istanbul in 1453 and the final demise of eastern half of the Roman Empire. That simple summary hardly captures the complexity of this movement; its ramifications are still being played out today. From the perspective of the history of science the important point is that Baghdad, Damascus, Cairo, Tehran, Cordoba, and other cities prospered and science flourished. These scholars, in contrast to their counterparts in China, were familiar with the writings of the Greeks and their translations to Arabic subsequently allowed medieval Europe its initial access to Greek scholarship.

Lord Dufferin in 1890 acknowledged: "It is to Mussulman science, to Mussulman art, and to Mussulman literature that Europe has been in a great measure indebted for its extrication from the darkness of the Middle Ages." C.H. Haskins wrote in 1927: "The broad fact remains that the Arabs of Spain were the principal source of the new learning for Western Europe." Arthur Glyn Leonard, 1909: "Do not we, who now consider ourselves on the topmost pinnacle ever reached by culture and civilization, recognize that, had it not been for the high culture, the civilization and intellectual, as the social splendors of the Arabs and soundness of their system, Europe would to this day have remained sunk in the darkness of ignorance?"

As is now well appreciated, the world of Islam had many achievements and insights beyond those of the Greeks. Yet, well before the expansion of European colonialism the science of Islam had run its course. As with China, one might ask the Needham question, *why?*

It is difficult to identify a single school or philosophical approach to science in Islam. There seemed to be an easy flow of men and ideas among these various centers — Bagdad, Damascus, Cairo, Tehran, Cordoba — and little conflict with, or regulation by, mosque or state. Summary descriptions of their various achievements speak to the strength of Islamic science, especially medicine, by 1300.

Al-Khawarizmi (780–850) was one of the founders of algebra; he employed a primitive form of logarithms. Al-Hasan b. al-Haitham (Alhazen) (965–1040) wrote works of mathematics and astronomy. He recorded eclipses, referred to the constituent colors of white light as seen in rainbows and invented a pinhole camera. He argued — contrary, to Plato, Euclid, and Ptolemy — that vision does not involve the emission of something from the eye. Al-Biruni (973–1050) suggested that the Earth rotates about its own axis. Shareef al-Idrisi (1100–1166) wrote of cartography and geography. Alhazen (Ibn al-Haytham) (965–1040) in his *Book of Optics* advocated experiments to test theories.

Abu Musa Jābir ibn Hayyān (~721–~815), known in the West as Geber, was a physician. He wrote a series of "books" or pamphlets, e.g. *Al-Zuhra* (*Book of Venus*) and *Kitab al-Ahjar* (*Book of Stones*) and the *Emerald Tablet* dealing with chemistry. He described the dissolution of gold in aqua regia (HCl and HNO_3) and reactions involving citric acid, acetic acid, tartaric acid, arsenic, antimony, and bismuth. As with many works, East and West, treating with alchemy, a bit of mystic symbolism (cf 119/11B5) seemed inherent to the protocols, hence the term "gibberish" (Chapter B5). In *Kitab al-Ahjar*, (*The Book of Stones*) he wrote "... alchemy is possible only by subjugating oneself to the will of Allah." He searched for al-iksir (elixir = make possible), i.e. the philosopher's stone. He argued the commonly held belief that the ratio of sulfur to mercury used in preparing the metal gave rise to different metals. Geber and contemporaries had yet to formulate a clear definition and theory of elements. He wrote "The first essential in chemistry is that you should perform practical work and conduct experiments, for he who performs not practical work nor makes experiments will never attain the least degree of mastery." Alchemists in western Europe from ~1000 to ~1700 pursued similar goals. How and to what extent they communicated with their Muslim predecessors is not well documented (Chapter B5).

Avicenna (Abū Alī al-usayn ibn Abd Allāh ibn Sīnā (980–1037) wrote *The Canon of Medicine and The Book of Healing*, in which he discussed systematic experimentation, contagion, and quarantine, mediastinitis (mid-chest membrane inflammation) vs. pleurisy (membrane surrounding the lungs), tuberculosis, dermatitis, and sexually transmitted diseases. He believed the application of leeches to be more useful than cupping in "... letting of the blood from deeper parts of the body." The use of zarnab, a mixture of alkaloids from *Taxus baccata*, for cardiac pain is the first recorded calcium channel blocker (Chapter C10). Several of his admonitions have a contemporary ring:

> The drug must be free from any extraneous accidental quality. The drug must be tested with two contrary types of diseases, because sometimes a drug cures one disease by its essential qualities and another by its accidental ones. The quality of the drug must correspond to the strength of the disease. For example, there are some drugs whose heat is less than the coldness of certain diseases, so that they would have no effect on them. The time of action must be observed, so that essence and accident are not confused. The effect of the drug must be seen to occur constantly or in many cases, for if this did not happen, it was an accidental effect. The experimentation must be done with the human body, for testing a drug on a lion or a horse might not prove anything about its effect on man.

His guidelines are reasonable, clear, and still relevant today; even though couched in the language of the time.

Numerous other authors wrote tracts on medicine: Ali ibn Sahl Rabban al-Tabari's Firdous al-Hikmah (~860), *Paradise of Wisdom*; Muhammad ibn Zakarīya Rāzi (Rhazes) (865–925), *The*

Diseases of Children, in which he was critical of humorism; Ali ibn Abbas al-Majusi (~980), *Kitab Kamil as-sina'a at-tibbiyya, Complete Book of the Medical Art.*

Ishaq bin Ali al-Rahwi (854–931) al-Raha, in Syria, referred to medical peer review in *Ethics of the Physician.* Ishaq bin Ali Rahawi, Adab al-Tabib (in the 800s), *Conduct of Physicians,* called practitioners "guardians of souls and bodies." *Kitab al-Saydalah* in *The Book of Drugs* gave details of drugs and duties of the pharmacist.

Abu al-Qasim al-Zahrawi (Abulcasis) (~1000), *Kitab al-Tasrif* (30-volume *Book of Concessions*), described forceps for use in childbirth. Ibn al-Thahabi (~1000), in his alphabetical encyclopedia of medicine discussed diabetes mellitus, "... describing the abnormal appetite and the collapse of sexual functions and he documented the sweet taste of diabetic urine."

Ali ibn Abbas al-Majusi (1000), proved false the view that the "fetus swims out of womb" (per Hippocrates, Galen, Ptolemy), showing instead that it is aided by uterine contractions. Abu al-Qasim al-Zahrawi (1000), *Al-Tasrif,* wrote on obstetrics and mentioned forceps, catgut sutures, ligatures, surgical needles, scalpels, curettes, retractors, the surgical spoon, surgical hook, surgical rod, specula, and bone saw. Ferdowsi (1010), *Shahnameh* and al-Biruni *Al-Athar al-Baliyah,* described a caesarean delivery. Ibn al-Haytham (1021), in the *Book of Optics,* discussed the role of the retina in perception. Constantinus Africanus (~1087), in Salerno wrote a textbook of *Schola Medica Salernitana.* Avenzoar (1091–1161), in Andalusia, argued that scabies is caused by a parasite (the mite, *Sarcoptes scabiei*), not by humorism.

Ibn al-Nafis (1213–1288) had a broad understanding of anatomy: "The permeation of arteries into the cranium is well known not to be from the front ventricle." "The most important muscles of a human body total 529 ..." He identified 10 cranial nerves and, in his *Book on Experimental Ophthalmology,* distinguished the muscles of eyeball from the optic nerve and noted that "… each nerve (of the eye) goes to the opposite side." He wrote that cognition, sensation, imagination, and locomotion emanate from the brain and described a harmony between religion and philosophy.

Muhammad ibn Zakarīya Rāzi (d. 925) wrote *Doubts on Galen*. He distinguished venous blood, which is dark, from arterial blood, which is light in color. He argued that the former came from the liver and the later from the heart and suggested that the "… blood pulse back and forth like tides." This view of circulation, analogous to respiration, was accepted until William Harvey published *De Motu Cordis* in 1628 (Chapter C3). Rāzi might be considered to be the first epidemiologist; he hung raw meat in various streets in Baghdad and sited his hospital in the area where the meat rotted least. Ibn al-Nafis practiced at the Medical College Hospital (Bimaristan al-Noori), Damascus, and at the Al-Nassri Hospital, 1236, and at Al-Mansouri Hospital, Cairo, as Chief of Physicians. He wrote the 80-volume *The Comprehensive Book on Medicine* in which he denied the existence of pores through the inter-ventricular septum. He wrote that blood from the right ventricle goes to the lungs; the lighter parts filter into the pulmonary vein, mix with air, and the blood mixed with air in the lung returns to the left ventricle, which receives nourishment from blood in vessels in its substance. He presented an accurate theory of pulsation. He elaborated: "The lungs are composed of parts, one of which is the bronchi; the second, the branches of the arteria venosa; and the third, the branches of the vena arteriosa, all of them connected by loose porous flesh."

> The primary function of contraction of the heart is to absorb the cool air and expel the wastes of the spirit and the warm air; however, the ventricle of the heart is wide. Moreover, when it expands it is not possible for it to absorb air until it is full, for that would then ruin the temperament of the spirit, its substance and texture, as well as the temperament of the heart. Thus, the heart is necessarily forced to complete its fill by absorbing the spirit. (Chapter C3)

Ibn al-Nafis noted that "… neither of the two semen has in it an active faculty to fashion"; they combine in the womb. Ibn al-Quff (1233–1286), a student of Ibn al-Nafis, described the formation of a foam stage in the first six to seven days, which in 13 to 16 days is gradually transformed into a clot and in 28 to 30 days into a small chunk of meat. In 38 to 40 days, the head appears separate from the shoulders and limbs. "The brain and heart followed by the liver are formed before other organs. The fetus takes its food from the mother in order to grow and to replenish what it discards or loses … There are three membranes covering and protecting the fetus, of which the first connects arteries and veins with those in the mother's womb through the umbilical cord. The veins pass food for the nourishment of the fetus, while the arteries transmit air. By the end of seven months, all organs are complete … After delivery the baby's umbilical cord is cut at a distance of four fingers breadth from the body, and is tied with fine, soft woolen twine. The area of the cut is covered with a filament moistened in olive oil over which a styptic to prevent bleeding is sprinkled … After delivery, the baby is nursed by his mother whose milk is the best. Then the midwife puts the baby to sleep in a darkened quiet room … Nursing the baby is performed two to three times daily. Before nursing, the mother's breast should be squeezed out two or three times to get rid of the milk near the nipple."

al-Quff continued: "Galen believes that each of the two semen has in it the active faculty to fashion and the passive faculty to be fashioned, however the active faculty is stronger in the male semen while passive in the female semen. The investigators amongst the falasifa (group of philosophers) believe that the male semen only has the active faculty, while the female only has the passive faculty … As for our opinion on this, and God knows best, neither of the two semen has in it an active faculty to fashion." "… once the male semen and female semen are brought together in the womb, the female semen quenches the hot fire of the male semen through its own cool and wet nature." (Chapter C5)

In the 1100s, two female physicians of the Banu Zuhr family served Almohad, ruler Abu Yusuf Ya'qub al-Mansur Şerafeddin Sabuncuoğlus. They dealt with not only clinical medicine but also fundamental questions of physiology and embryology. One may explore the sophistication of each of these practitioners. However, the general conclusion is solid: the practice of critical, informed medicine was widespread in the Islamic world. What then led to its stasis well before European conquests? This is a complex story, well beyond the scope of this book; however the Arab world was well aware of science in Europe. Darwin's *Origin* (Chapter C14) was translated to Arabic and had an impact on their scholarship and faith (Elshakry M. *Reading Darwin in Arabic, 1860–1950*).

A5

Christianity and Science

St. Thomas Equinas as depicted in *Wisdom Conquers Evil*
(Santa Maria Sopra Minerva).

In contrast to the triumphant spread of Islam, the early centuries of Christianity were tenuous. The gospels of Mark, Matthew, Luke, and John, as well as of the early convert, Paul, gave different accounts of the life and teachings of Christ. Scores of heresies were debated and condemned. This complex web of interactions played out not only in the Roman province of Judea but throughout the Empire.

From the perspective of the development of science, there were two major themes: The early Church leaders, especially Paul (5–67 A.D.) and Augustine of Hippo (354–430), rightly argued that if the Church were to survive, it must establish a single doctrine and discipline its converts. This meant that philosophizing and speculation in matters both of faith and of science were not to be tolerated. Paul was unequivocal: "The more they called themselves philosophers, the more stupid they grew (Romans 1:21–22)." He declared war on the Greek rational tradition through his attacks on "... the wisdom of the wise ..." and "... the empty logic of the philosophers."

Pope Gregory the Great warned those with a rational turn of mind that, by looking for cause and effect in the natural world, they were ignoring the cause of all things, the will of God. Science in the Greek world under the nominal suzerainty of the emperors in Rome and in Constantinople declined during the first four centuries A.D. This decline paralleled the rise in power and intolerance of the Church in both the western and eastern empires. Whether this correlation reflects causality remains a matter of debate.

The Draper–White thesis "… wedded a triumphalist view of science with a patronizing view of religion." "Grounded in faith, religion seemed bound to suffer when confronted by science, which was, of course, based on fact." — Gary Ferngren in *Science and Religion* (2002). These two books — *History of the Conflict between Religion and Science* (1874) by John William Draper, and *A History of the Warfare of Science with Theology in Christendom* (1896) by Andrew Dickson White — informed American views of the supposed conflict between religion and science.

E.R. Dodds in *The Greeks and the Irrational* (1951) wrote "… honest distinction between what is knowable and what is not appears again and again in fifth-century (B.C.) thought, and is surely one of its chief glories."

Colin Russell conceded "While it cannot be denied that isolated cases of real conflict have existed, as in the cases of Galileo and Darwin, recent historiography suggests that it would be wrong to extrapolate from these examples to the view that science and religion are necessarily hostile." Ferngren continued "… Christianity has often nurtured and encouraged scientific endeavor, while at other times the two have co-existed without either tension or attempts at harmonization. The story of science and Christianity in the Middle Ages is not a story of suppression nor one of its polar opposite, support and encouragement."

Charles Freeman in *The Closing of the Western Mind: The Rise of Faith and the Fall of Reason* (2005) argued that "… the central theme of this book (is that) … the Greek intellectual tradition was suppressed rather than simply faded away." He continued: "Paul … declared war on the Greek rational tradition through his attacks on "… the wisdom of the wise" and on "the empty logic of the philosophers." Paul "… formulated a meaning for Jesus' death and resurrection." "Unlike Jesus he insisted on a dramatic break with traditional culture, not only his own, but also that of the Greco-Roman world." He wrote "... that is anyone preaches a version of the Good News different from the one we have already preached to you, whether it be ourselves or an angel from heaven, he is to be condemned." "... an exploration of its (the crucifixion) meaning forms the core of his theology." "... sin is a heavy, albeit abstract, entity that burdens the human race."

Freeman continued: "Whereas in traditional Greco-Roman religion the public observation of rituals is primary, Paul presents something radically different proposing that the orientation of the inner person to God is essential. It is an idea that reached fruition in Augustine, who in his *Confessions*, talks of God actually being inside a person's intimate being and in a continual and often in Augustine's case, stormy relationship with him." "In the second Letter to the Thessalonians it is made clear that those who refuse to accept 'the Good News of our Lord Jesus Christ' will be punished for eternity (1:9)." "So for Paul it is not only the Law that has been superseded by the coming Christ, it is the concept of rational argument, the core of the Greek intellectual achievement itself." Paul preached "The more they (non-Christians) called themselves philosophers, the more stupid they grew (Romans 1:21–22)." "Gentile Christianity, through Paul, had declared war on the Greco-Roman world, its gods, its idols and its mores." "It was as a result of the urgent need to define its boundaries and beliefs that Christianity developed sophisticated notions and structures of authority."

Freeman concluded: "By fixing on a comprehensible symbol, the death and resurrection of Christ, by proclaiming the enormous and imminent rewards of Christian faith and the awful consequences of rejection of 'the cross of Christ', Paul had created a focus for community worship." "So here are the roots of the conflict between religion and science that still pervades debates on Christianity to this day."

The Roman emperor Diocletian in 302 allied himself with the ancient gods seemingly to minimize conflict:

> The immortal gods in their providence have so designed things that good and true principles have been established by the wisdom and the deliberation of eminent, wise and upright men. It is wrong to oppose these principles, or desert the ancient religion for some new one, for it is the height of criminality to try and revise doctrines that were settled once and for all by the ancients, and whose position is fixed and acknowledged.

Constantine defeated, then executed his rival in the East, Licinius, in 325; Licinius II was killed in 326. Constantine was supreme within the empire and, sitting on the "throne of gold," convened the bishops in the first council of Nicaea in 325. In all modesty he pronounced: "We have received from Divine Providence the supreme favor of being relieved from all error." Constantine was finally baptized in 337, weeks before his death. Thus began the jostling for state control of the Church and the Church's control of the state. The reverberations of this competition are still being played out in some countries today. Jesus became a god of war when about 375, Emperor Ambrose in *De Fide* proclaimed "... the army is led not by military eagles or the flight of birds but by your name, Lord Jesus, and Your Worship." This illustrates the fundamental significance of separation of church and state in many contemporary industrialized nations.

In 378, Goth refugees, fleeing the Huns, crossed the Danube. In the Battle of Adrianople, the Huns defeated Emperor Valens; 10,000 Roman troops were killed. Later, Theodosius, the local commander, ordered a massacre in 387 in retaliation for a rebellion in Thessaloniki. He asked for penance from Ambrose, recently baptized and installed as bishop, thereby establishing the precedent of the Church's forgiving and legitimizing political authority. Rome was sacked by the Visigoths in 410. This was the inflection point in the fall of the Western Empire and beginning of the Dark Ages.

Freeman continued: "Augustine believed that every other form of learning had to be subordinated to the scriptures ... secular knowledge, whether provided by mathematicians, scientists or philosophers, is said to be valid only insofar as it leads to an understanding of scripture." Even more extreme were "... John Chrysostom's exhortations to Christians to empty their minds of secular knowledge." "One important theme which has run though this book is the linking of belief in rational thought with a belief in free will."

"Thomas Aquinas (1225–1274) revived the Aristotelian approach to knowing things so successfully that he unwittingly laid the foundations of the scientific revolution that was to transform western thought." "In the year of his breakdown (1273) he was strongly criticized in Paris for his insistence on a natural underlying order of things" "It is not until the fourth and final book of the *Summa Contra Gentiles* that he introduces those Christian doctrines sustainable only by faith, which he includes the doctrine of the Trinity, the Incarnation and the creation of the world by God, *ex nihilo* ... (The alternative view, held by both Aristotle and Plato (Chapter A2), which Aquinas accepted he could not disprove, was that matter had existed eternally alongside God)."

One is left to ponder whether Europe's millennium of darkness, from the sack of Rome in 410 to the siege of Vienna in 1529, should be attributed to the westward pressures of tribes from Asia, or to the anti-intellectual policies of the early church. As is often the case in historical analysis, one can document the occurrence of two events, e.g. the intolerance of early Christianity and the decline of science in Europe. The inference of causality is more tenuous, and yet ...

A6

Inductive Logic, "Works," and Francis Bacon

Francis Bacon
(1561–1626).

Francis Bacon was one of many who tried to free Western thinking from simplistic (mis)interpretations of Aristotle's works, or so-called scholasticism. He is credited with explicitly advocating what we now refer to as the inductive method. That is, one makes a series of objective observations, or experiments, and from them finds correlates. Many critics have noted the obvious: correlation does not prove causality. However, such associations underlie much of our learning and behavior. Later we will describe a more refined synthesis of inductive and deductive logic.

Fully as important, Bacon argued the societal value of "works" — what we might now call applied science — engineering and medicine. He also urged the government, King and parliament, to support scholars pursuing these works. Subsequently, many scientists, correctly or not, argued that they were following Bacon; he was cited by the founders of the *Royal Society of London for the Improvement of Natural Knowledge* in 1642. Whether his greatest influence lay in elaborating the inductive method or in his championing the value of works is problematic. The Scientific Revolution nominally occurred between 1600 and 1700. Bacon was one of its more influential figures even though he was not a scientist.

Francis Bacon (1561–1626) was the youngest of five sons of Sir Nicholas Bacon, Lord Keeper of the Great Seal for Elizabeth I. Francis and an older brother attended Trinity College, Cambridge from 1573–1576; he then traveled to Paris with his father who died in 1579 and left him only a small inheritance. Bacon attended Gray's Inn, residence in law, 1579, and became an outer barrister in 1582. He became a member of parliament in 1584 and urged the execution of Mary Queen of Scots. He was deeply involved in the intrigues of Elizabeth's reign and was an advisor to Elizabeth's one-time favorite, Robert Devereux, second Earl of Essex. Bacon subsequently investigated charges of treason against Essex and pressed for his execution in 1601.

Bacon was knighted by James I in 1603 and assumed the clerkship of the Star Chamber in 1608. He wrote the government report "The Virginia Colony" in 1609 and helped form the Newfoundland Colonization Co. in 1610. In 1618 he was charged with corruption (debt) by the Lord Chancellor and debarred in 1621.

Quite remarkably, Bacon had the breadth of interest and the energy to write, in several versions, the *Advancement of Learning* (1605), *De sapientia veterum* (*Wisdom of the Ancients*, 1609), *Novum organum* (1620), and the *New Atlantis* (1626). He criticized the pedantry of "post-Aristotelians" and described utopias: *Christianopolis*, Andreae (1619), *City of the Sun*, Campanella (1623). He argued that "... the sciences in their present state are useless for the discovery of works, so logic in its present state, especially obsession with 'Aristotelian' syllogisms, is useless for the discovery of sciences."

Bacon felt that the acquisition of knowledge must precede generalization and focused on the arrangements of things previously discovered, not methods of discovery. "... the sciences we now have are no more than elegant arrangements of things previously discovered, not methods of discovery or pointers to new results."

"... current logic is good for establishing and fixing errors (which are themselves based on common notions) rather than for inquiring into truth." He rejected "induction by enumeration" and advocated "rejections and exclusion" to trim conclusions down to one (as might be done by a lawyer or administrator). "The end of induction is the discovery of forms, the ways in which natural phenomena occur, the causes from which they proceed." Bacon had limited talent in mathematics and little sympathy for abstract theorizing.

He described an ideal college, "Solomon's House," remarkably similar to a modern research university devoted to applied and pure science. "The true and legitimate goal of the sciences is to endow human life with new discoveries and resources." "Just let man recover the right over nature which belongs to him by God's gift, and give it scope; right reason and sound religion will govern its use." These were appropriate Christian purposes. In his *The Essays: Of Atheism*, "... a little philosophy inclineth man's mind to atheism; but depth in philosophy bringeth men's minds about to religion."

These new ways of thinking first required rejection of:

- "Idols of the Tribe" (*idola tribus*), which are common to the race;
- "Idols of the Den" (*idola specus*), which are peculiar to the individual;
- "Idols of the Marketplace" (*idola fori*), coming from the misuse of language;
- "Idols of the Theatre" (*idola theatri*), which result from an abuse of authority.

Bacon was critical of contemporary (1600) interpretations of antiquity. "For the discovery of things is to be taken from the light of nature, not recovered from the shadows of antiquity." "... after Socrates had brought philosophy down from heaven to earth, moral philosophy grew still stronger, and turned men's minds away from natural philosophy." John Locke was influenced by Bacon and argued in his *Essay Concerning Human Understanding* (1690) that empiricism was the initial route to knowledge. For the next century, philosophers on the continent were more inclined to theoretical constructs than were their brethren in Britain.

Bacon was aware of the advanced state of science in China (Chapter A3). "Printing, gunpowder and the compass: These three have changed the whole face and state of things throughout the world; the first in literature, the second in warfare, the third in navigation; whence have followed innumerable changes, in so much that no empire, no sect, no star seems to have exerted greater power and influence in human affairs than these mechanical discoveries." Catching up was seen as a goal for Europe.

He understood that observation alone did not constitute inductive logic and offered a parable. Ants (as empiricists) "only collect and use"; spiders (rationalists) "make cobwebs of their own substance"; however, the bee "gathers ... transforms and digests it by a power of its own."

Peter Dear, in *Revolutionizing the Sciences* (2004), paid him the highest compliment: "The modern world is much like the world envisaged by Francis Bacon."

A7

Deductive Logic, Maths, and René Descartes

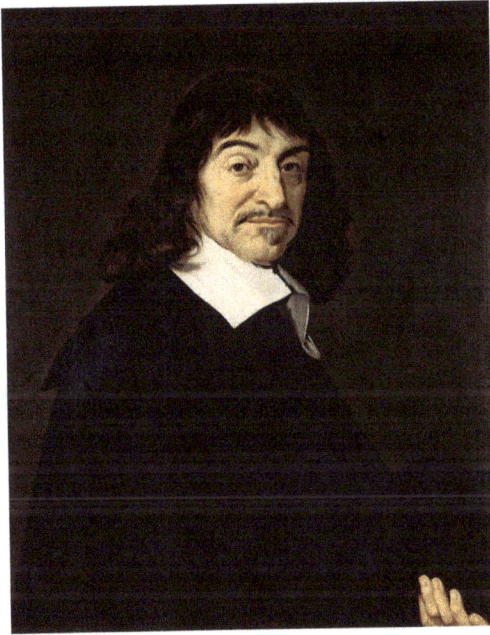

René Descartes
(1596–1650).

René Descartes, for whom the Cartesian coordinate system is named, is credited with describing and refining deductive logic, as summarized in his *Discourse on the Method* (1637). That is, given an initial set of assumptions and their logical development, the conclusion must be true if the assumptions were correct and the development was free of error. Although this assertion is generally accepted, it begs the question of how one chooses those initial assumptions. This deductive approach is often presented as the essence of how mathematics and physics is done. Yet, getting those initial assumptions and making legitimate approximations during the analysis can be extremely challenging and ultimately is informed by insights that need not be entirely logical.

Descartes championed dualism, i.e. the body works like a machine and obeys laws of physics; the mind does not. This is an extension of the arguments of Plato. He is remembered for the quotation "*Cogito ergo sum*" — "I think, therefore I am." His writings were generally endorsed by rationalists — Spinoza, Pascal, Leibniz — on the continent and criticized by empiricists — Hobbes, Locke, Berkeley, and Hume — in Britain.

Very roughly, this argument between rationalists and empiricists can be seen as an extension of alternate views of Plato and of Aristotle, captured in *The School of Athens* by Raphael, with Plato pointing to the heavens for mathematical purity and Aristotle pointing to the earth for observation.

René Descartes was born in La Haye en Tourain, near Richelieu, in 1596 and succumbed to the cold in Stockholm in 1650. He entered the Jesuit Collège, Royal Henry-Le-Grand, La Flèche in 1607 and received his baccalauréat and license in law in 1616 from the University of Poitier. His family was not wealthy; he had to work for his living. He served as secretary to Maurice of Nassau, leader of the United Provinces, Netherlands, in 1618 and was present at the siege of La Rochelle by Cardinal Richelieu in 1627. He visited or attended the universities of Franeker, 1629; Leiden, 1630; and Utrecht, 1635.

Descartes developed analytic geometry based on a coordinate system with three orthogonal axes, created exponential notation, and applied infinitesimal calculus to the tangent line problem. He helped wed the geometry of the Greeks to the algebra of the Hindus (Chapter B2).

From the law of refraction, ($n = \sin i/\sin r$; $n =$ index of refraction, $i =$ angle of incidence, $n =$ angle of refraction), he deduced the angular radius of a rainbow, 42°. He proposed the conservation of momentum. However, not all of his science was so analytical; he regarded the pineal gland as "the seat of the soul."

His *Discourse on the Method* (1637) was an explicit exposition of his deductive logic. Descartes maintained that the senses lack certainty; they can deceive. Human reason, including mathematics, is subject to error. He sought a fundamental set of principles that one can know to be true without any doubt. He wanted certainty rather than mere opinion; his ideas were to be "... accepted for their truth, not simply for their likelihood or even mere ingenuity." He sought to explore logic "... as if no one had written on these matters before." Descartes questioned what one actually senses: "Thus what I thought I had seen with my eyes, I actually grasped solely with the faculty of judgment, which is in my mind." He championed dualism — that is, the body works like a machine and obeys laws of physics; the mind does not. He asked, as we might today, how a non-material mind, as opposed to a brain, can influence a material body. He rejected appeal to ends, divine or natural, in explaining nature and proposed to proceed not "... as others usually do by way of aimless and blind enquires and more by luck than by skill but by following certain rules."

His proof of the existence of God may seem a bit convoluted to the modern ear. Man "... being imperfect, could have acquired the concept of perfection only from the perfect," that is, from God. Descartes argued that analysis by the mind is the essence of being. "*Cogito ergo sum*" — "I think, therefore I am."

He turned to science; all matter was assumed to be inert. This meant that a piece of matter had no propensity for moving itself — it was, in effect, dead. Thus the only way to get it to do anything was to apply to it some outside moving agency. Descartes rejected Aristotle's analysis that a cloth is red because it possesses redness; a fire is hot because it has warmth. He criticized these mere psychological impressions; one must look to the inherent qualities of matter. The only true idea of the nature of a body is geometrical extension of what matter really, in itself, is.

Descartes equated space with matter; hence there could be no vacuum, a view originally championed by Parmenides (Chapter A2). He attributed a straight, unbending flow to tiny particles thereby conserving momentum and a driving force on up to planetary motion. The "... action of light ..." should "... operate in straight lines emerging from the luminous body." "... the water that they (fish) push before them does not push all the water in the pool indiscriminately; it pushes only the water which can best serve to perfect the circle of their movement and to occupy the place which they vacate." The sun is an appearance generated by the presence at the center of our system of matter that consists of especially small, fluid, and very rapidly moving particles; their jostling exerts pressure." "... it is not surprising that the particles of salt have a sharp and penetrating taste, which differs a great deal from that of fresh water; because they cannot be bent by the fine material that surrounds them, they must always enter rigidly into the pores of the tongue ..."

Peter Dear wrote that in "Descartes' view of the natural world ... math-operational form of knowledge was capable of discussing, and no more." "Descartes wanted to present explanations that could not (he hoped) possibly be challenged. In other words, he wanted certainty rather than mere opinion; his ideas were to be accepted for their truth, not simply for their likelihood or even mere ingenuity." (*The Intelligibility of Nature: How Science Makes Sense of the World* (2006))

Like those of the Greeks and many others who preceded him (Chapter A2), many of Descartes' explanations or models of science were "wrong," given our contemporary understanding. One might ponder which of our models will be considered wrong in 2100 A.D. However, as will be elaborated, the important point is that he presented specific, reasonable ideas — the starting point for their replacement by better ones.

A8

The Scientific Revolution

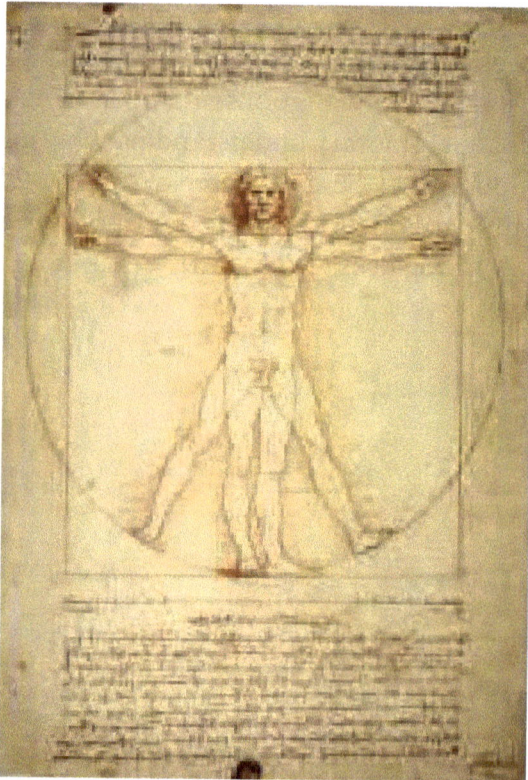

Leonardo da Vinci's
"Vitruvian Man"

(Vitruvius ~50 B.C.
De architectura).

It is misleading to speak of "the" scientific revolution. Inflection points of insight occurred at different dates in different fields and in different countries of Europe; the sequence of chapters in Sections B (physics) and C (biology) to some extent reflect these different dates. Certainly 1543 is the reference for both physics, Copernicus, and for biology, Vesalius. If that were not enough, Pierre Ramus in 1543 published his *Animadversions on Aristotle*. Not only must the new be elaborated, the old must be revised.

Many arguments are, at core, semantic; this makes them no less important, but it does shift the discussion. Since 1600, frequently cited as the beginning of the scientific revolution, a term coined by Alexandre Koyré, the rate of discovery has increased and continues to increase. This begs the question of the significance of those discoveries, applications, or insights. Yet, asking this rate may pose the wrong question. Did this period mark a fundamental difference in the way Europeans viewed themselves and their physical environment? If so, was the revolution more cultural than scientific? Did the increased rate of discovery reflect a change in culture? If so, why did China or Islam not experience such a revolution? To anticipate Thomas Kuhn's *The Structure of Scientific Revolutions* (Chapter A11), were there exceptionally many or major shifts of paradigm during this century ($\sim$1600–$\sim$1700)?

Certainly the breadth of economic and intellectual activities expanded. Gutenberg's invention of a printing press with movable type ($\sim$1440) facilitated the distribution of the 95 theses (posted 1517) of Martin Luther.

Nation states such as France, Britain, Spain, Sweden, and the Netherlands acquired greater stability and the peaceful transfer of power from one ruler to the next was more assured. One might argue that some of these changes were driven by advances in engineering, e.g. ship building, but hardly by basic science. More persuasive, perhaps, is the inverse argument that noblemen and merchants provided support for scientists, as championed by Francis Bacon in his Solomon's house described in *New Atlantis* (1626, Chapter A6).

There are three, at least, rather distinct ways to think about changes in science. The first and the second deal with progression, and with progress, from qualitative and quantitative perspectives.

Though difficult in practice, one could in concept list and weight the scientific activities of each decade and plot the weighted sum as a function of time. Certainly from 1300 onward such a graph for science in Europe would have a positive second derivative. A revolution in science, or within a sub-discipline, might be identified by a significant change in slope. The third approach, to be discussed in Section D, deals with the more complex interactions of society and science.

A more nuanced approach might adopt a Kuhnian perspective and ask which of the scientific activities on the list contributed to a shift in paradigm, as opposed to the accumulation of more information within an existing framework, i.e. ordinary science.

By all three metrics science, especially physics, from 1600 to 1700 experienced a revolution. This chapter summarizes a few of those key events; their details will be discussed in subsequent chapters.

Herbert Butterfield (1900–1979) in *Origins of Modern Science* (1949) wrote that The Scientific Revolution "… outshines everything since the rise of Christianity and reduces the Renaissance and Reformation to the rank of mere episodes … [It is] the real origin both of the modern world and of the modern mentality. Alexandre Koyré (1892–1964) in *From the Closed World to the Infinite Universe* wrote in 1959 of "… the most profound revolution achieved or suffered by the human mind … since Greek antiquity." P. Williams and H.J. Steffens, in *The History of Science in Western Civilization* (3 Vols. 1978–79), wrote: "It is a unique event, having never occurred in any other place or time and its effect on the development of Western civilization ranks it among the greatest events of human history." Steven Shapin in 1996 offered a catchy introduction: "There was no such thing as the Scientific Revolution and this is a book about it," but he basically agrees.

Williams and Steffens continued that scholars left scholasticism, the barren study of a simplified Aristotle, and scientists began to study Nature "in the raw." They rediscovered Plato and redefined physical reality in terms of mathematics. Many components of society, from the Mediterranean to Britain, were engaging a more demanding and more rewarding world. Four of the major factors that laid the foundation for the scientific revolution were universities, exploration, the Renaissance, and the Reformation.

Many universities were established in Europe before or during the 1200s; the approximate dates are:

Bologna   1088
Paris   1150
Oxford   1167
Modena   1175
Palencia   1208
Cambridge   1209

Salamanca	1218
Montpellier	1220
Padua	1222
Toulouse	1229
Orleans	1235
Siena	1240

Some were administered and supported by the Church, others by student organization; Oxford and Cambridge were supported by the Crown. The early faculties addressed the "trivirium" of grammar, logic, and rhetoric; subsequently they added the "quadvirium" of arithmetic, geometry, astronomy, and music.

Students entered as young as age 14; about six years study led to a Master's degree. Faculties of law, medicine, and theology were added; a doctorate might be awarded after about eight years. Rediscovered Greek manuscripts, often translated to Arabic, were brought to the West, especially by scholars fleeing the fall of Constantinople in 1453. Although the curricula were limited and the study often focused on a corrupted Aristotelian scholasticism, these early universities shared common features and curricula; students and faculty could travel from one to another and feel at home. They survive to this day and evolved to set a rather standard pattern for universities throughout Europe, then the world.

The Renaissance is roughly dated ~1500–~1600; the scientific revolution, ~1600–~1700. Whatever spans are chosen, the important point is that the Renaissance preceded the revolution. In one sense the Renaissance respected Greek scholarship; the revolution rejected much of it. One is left to wonder in what ways the Renaissance affected science?

There were several renaissances; the fever spread north from Italy. Giotto (1267–1337) employed a sort of linear perspective and achieved a natural reality in painting unknown to his contemporaries. Dante Alighieri (1265–1321) in Florence wrote in the vernacular to an increasingly literate citizenry. Petrarch (1304–1374), perhaps the leading intellect of his time, revisited classical manuscripts. Filippo Brunelleschi (1377–1446) and Alberti (1404–1472) drew and wrote sophisticated works about architecture.

Merchants of Florence, for example Lorenzo de' Medici, developed complex practices of finance and commerce; they were quite anti-monarchical. Many became patrons of the arts. One still marvels today at the brilliance of Leonardo da Vinci (1452–1519) and Michelangelo (1475–1564), appropriately designated "renaissance men."

Pico della Mirandola wrote of human dignity, *De hominis dignitate* (1486). Niccolò Machiavelli (1469–1527) and Thomas More (1478–1535) contemporary government. In 1543, both Copernicus in *De Revolutionibus* (Chapter B3) and Andreas Vesalius in *De humani corporis fabrica* (Chapter C2) questioned the received wisdom of Ptolemy and of Galen, respectively.

This was the age of discovery. There had been voyages to Somalia by the Chinese in the 800s (Chapter A3); the Vikings first settled Iceland in 865. Marco Polo returned to Venice in 1271 to reveal his, and his father's, adventures; the Europeans were aware of the fabulous spice islands. In 1330, Jodanus de Severec established the first French missions in India. In 1402, Bethencourt founded the first European settlement in the Canary Islands, the launching point for Columbus' voyage to the "Indies" in 1492. In 1515, Balboa (1475–1519) walked over Panama to view the Pacific. Magellan (1480–1522) was killed in the Philippines; only *Victoria* of his five ships completed the first circumnavigation of the globe. The horizons of Europeans had been expanded, as never before and, one might argue, as never since.

Perhaps more important, their internal horizons were changing. Working in both Strasbourg and Mainz, Johannes Gutenberg (~1398–1468) developed a printing press with movable type about 1439. He printed many copies of his Gutenberg Bible; perhaps more important, his press could produce lots of cheap pamphlets. Several questioned practices of the Church. How would one compare the relative impacts of Gutenberg and Google?

There were many conflicts within the Catholic Church. For example, the Albigensian Crusade (1209–1229) suppressed Catharism in Languedoc. However, this was not a direct antecedent of the Protestant reformation. John Wycliffe (1320–1384) received this doctorate in theology and became master of Balliol College, Oxford, in 1361. He was critical of the Church's accumulation of wealth as elaborated in his 18 theses and *Summa Theologiciae*. He translated the Bible into vernacular English in 1382. His open criticism of the Church served as a model for subsequent reformers. Jan Hus (1372–1415) of the Moravian Brethren was promised safe passage to the Council of Constance, then burned at the stake. The Council ordered that Wycliffe's body be exhumed and burned posthumously.

Martin Luther (1483–1546) had preached against the sale of indulgences and against simony, the purchase of positions within the Church. He sent his 95 theses (*Disputation of Martin Luther on the Power and Efficacy of Indulgences*) in 1517 to his bishop, Albert of Mainz. He protested in 1520: "Hier stehe Ich, Ich kann nicht anders" (Here I stand; I cannot do otherwise) and was excommunicated in 1521. The Diet of Worms, 1521, forbade anyone from this time forward to dare, either by words or by deeds, to receive, defend, sustain, or favor Martin Luther.

Each of the following — Huldrych Zwingli, executed in Zurich; John Calvin, who moved from Paris to Geneva; and John Knox of Scotland — had his own message of reformation. There was hardly a united front against the Catholic Church. The Church of England separated from Rome in the period 1529–1536. Thomas More stood opposed and was beheaded by order of Henry VIII of England. The Council of Trent (1545–1563) condemned the Protestant heresies. About 50,000 Huguenots were killed in Paris in the St. Bartholomew's Day massacre, 1572. Henry IV of France granted some religious freedom to the Huguenots in the Edict of Nantes, 1598. These rights were revoked by the Edict of Fontainebleau, 1685.

In the Thirty Years' War (1618–1648), the Catholic House of Habsburg, supported by Spain and Austria, fought the protestant princes of Germany, supported by Denmark, Sweden, and France. The slaughter killed about 35% of the German population. In the Treaty of Westphalia, 1648, all parties recognized the Peace of Augsburg, 1555; each prince had the right to determine the religion — Roman Catholicism, Lutheranism, or Calvinism — of his state. All guarantied the practice of other faiths in public during allotted hours and in private at will. Pope Innocent X was not amused and declared the treaty "… null, void, invalid, iniquitous, unjust, damnable, reprobate, inane, empty of meaning and effect for all times."

Europe was busy from 1500 to 1700. Important discoveries — changes in paradigm, to anticipate Kuhn's terminology — were made in science. The correlation is irrefutable; as always, the question is causality.

As will be elaborated in the following chapters, a new physics was being born. Copernicus (1473–1543) in *De revolutionibus orbium celestium* proposed a heliocentric system in 1543. Gilbert (1544–1603) discussed magnetism in *De magnete* (1600) and suggested a d^{-2} force relationship. Bacon (1561–1626), in the *Novum organum* (1620), discussed the inductive method, but more importantly championed a research institute devoted to applying "works" to the benefit of society (Chapter A6). Kepler (1571–1630) described the elliptical orbits of the planets and the ratio of the square of the period to the cube of the major axis of the orbits of two planets ($T_1^2/T_2^2 = r_1^3/r_2^3$). Galileo (1564–1642) observed four of the moons of Jupiter and recorded the movement of "spots" across the surface of the sun (Chapter B3). He noted that two masses fall at the same velocity under the same gravitational force in Pisa and supposedly worldwide. He combined the two rectilinear motions of Aristotle into the curved path of a cannon ball. Descartes (1596–1650) elaborated on the deductive method, developed analytical geometry, and derived the equation describing the trajectory of a projectile. Newton (1643–1727), in his *Principia* (1687), assumed that Gilbert's inverse square law applied to gravitational attraction: $F = k \cdot m_1 \cdot m_2 / d_{12}^2$ (force of attraction is proportional to the product of the two masses divided by the square of their distance apart). He proposed his three laws of motion: A body moving in a straight line continues straight until another force acts upon it. Acceleration equals the force acting upon a body divided by its mass: $F = m \cdot a$. For every action, there is an equal and opposite reaction. He developed a form of calculus, so-called fluxions, and from his three laws derived the orbits described by Kepler. Most importantly he proposed that these laws were universal (Chapter B4).

Biology experienced major advances: Vesalius (1514–1564) in *De humanis corporis fabrica* (1543) corrected several errors in Galen's anatomy. His work was followed by that of a series of distinguished anatomists, most working at the University in Padua, who explored human anatomy in ever finer detail. Harvey (1578–1657) studied at Padua and several other universities on the continent. In his *De motu Cordis* (1628) he correctly described the complete circulatory system. He found no hole in the septum of ventricles, or of the auricles; he postulated a capillary bed, subsequently visualized by Malpighi, that connects the arterial and venous systems (Chapter C3).

One might rephrase the question of the origin(s) of the scientific revolution. Thomas Carlyle (1795–1881) wrote: "Universal history, the history of what man has accomplished in this world, is at bottom the History of the Great Men who have worked here." In contrast, Otto von Bismark (1815–1898) wrote: "The statesman's task is to hear God's footsteps marching through history, and to try to catch on to His coattails as He marches past." Why did God choose to place his scientific steps in Western Europe at this time?

A9

The Church and Science

Campo de' Fiori (Rome)
Bruno's execution.

Committee: Hugo, Spencer, Renan, Haeckel, Ibsen and Gregorovius, 1885
opposed, clerical party; erected by Rome Municipality,1889

Survival was the main challenge for the early Christian Church. This depended on establishing a doctrine and suppressing heresy. The elaboration and articulation of doctrine was a significant challenge. Broad intellectual curiosity about science or other frivolities was not encouraged and often suppressed.

By ~600, the Church was a bit more secure. The bishop of Rome was called the Pope, at least in Western, Catholic Europe.

Inevitably members of the priesthood and general laity were curious about the natural world. These involvements did not threaten the Church; they were in general tolerated if not explicitly supported. By the new millennium, men like Roger Bacon and Thomas Aquinas were thinking more broadly; there was no conflict between their science and their Church. Modest advances in alchemy, metallurgy, and engineering were generally embraced. Monasteries were often centers of technical expertise, especially in agriculture and medicine. However, Luther's posting (apocryphal or not) of his 95 theses in 1517 did not make the Church more tolerant. Most of the distinguished scientists of the scientific revolution — da Vinci, Copernicus, Vesalius, Kepler, Brahe, Galileo, Harvey, Huygens, Descartes, Newton, Leibniz — expressed their commitment to the Church and the true faith.

The Draper–White thesis argued that for the preceding three centuries, 1600–1900, religion and science had been in conflict and science had inevitably won. Several contemporary historians have revised this triumphalist or whiggish interpretation. They concede the conflict between Galileo and Pope Leo X as well the opposition to Darwin and Wallace's theory of evolution; however, most religious authorities, most of the time have not only accepted but also championed science, ever more so in recent years. The 76 members, including several Nobel Laureates, of the Pontifical Academy (re-established in 1936) are committed to promoting science and to exploring the relevance of science to epistemological concerns. The (in)tolerance of the Church has not been constant over the centuries; both critics and defenders can choose from many examples.

The historian can select, and accurately portray, examples to support one interpretation of a complex story. The following examples reflect, hopefully, a balanced view of the evolving relationship between science and the Catholic Church.

Hildegard von Bingen (1098–1179), the tenth child of a family of free nobles, was "offered" to the church as a tithe. She was elected magistra by her fellow nuns and founded a second convent for her nuns at Eibingen in 1165. In *Causae et curae*, she described tinctures, herbs, precious stones and the cosmos; she concluded "... all things put on earth are for the use of humans."

Roger Bacon (∼1214–∼1294) was a Franciscan friar; he then became a Master at Oxford. Bacon recorded the spectrum from white light passed through a "prism" and opposed blind obedience to authority. His *Opus majus* and *Opus minus* included a critique of Aristotle and essays on mathematics, optics, alchemy, gunpowder, and astrology. He was applauded by his students as Doctor Mirabilis and was valued by colleagues and by the Church.

Andreas Osiander prepared an unauthorized preface to Copernicus' *De revolutionibus orbium coelestium* (On the Revolutions of the Heavenly Spheres, 1543); he cautiously noted that it did not reflect physical reality, but was a device for computing orbits (Chapter B3). Nonetheless, the Church in 1616 placed it on its *Index librorum prohibitorum* (List of prohibited books) and suspended *De revolutionibus* for correction; it was declared "... false and altogether opposed to the Holy Scripture." Paracelsus (1493–1541) wrote in his discussion of alchemy "This is the way that nature proceeds with us in God's creatures..."

Miguel Serveto Conesa (1511–1553) studied with Dominican friars and was the first European to discuss pulmonary circulation. He was condemned by the Catholic Church as a religious dissenter and anti-Trinitarian. Jean Calvin wrote: "Servetus has just sent me a long volume of his ravings. If I consent he will come here, but I will not give my word for if he comes here, if my authority is worth anything, I will never permit him to depart alive." He was arrested by Calvin and burned at the stake in Geneva. His conflict had nothing to do with pulmonary physiology; however, his interest in science hardly offered any protection.

Giordano Bruno (1548–1600) was ordained as a Dominican in 1572, then left the Order and was excommunicated. He championed his own mystical heliocentrism and a pantheistic hylozoistic system (all matter is living). He was arrested by the Church in 1592 and transferred to Rome next year. His trial lasted seven years. He was convicted of:

- Holding opinions contrary to the Catholic Faith and speaking against it and its ministers.
- Holding erroneous opinions about the Trinity, about Christ's divinity and Incarnation.
- Holding erroneous opinions about Christ.
- Holding erroneous opinions about Transubstantiation and Mass.
- Claiming the existence of a plurality of worlds and their eternity.
- Believing in metempsychosis and in the transmigration of the human soul into brutes.
- Dealing in magic and divination.
- Denying the Virginity of Mary.

Cardinal Bellarmine demanded full recantation; Bruno refused. Pope Clement VIII denied his appeal. Bruno defiantly responded: "Perhaps you, my judges, pronounce this sentence against me with greater fear than I receive it."

He was gagged, tied to a pole naked, and burned at the stake, in 1600, in Campo de' Fiori, Rome. All of his works were placed on the *Index* in 1603. John Paul II acknowledged the Church's error in 2000 and expressed "profound sorrow." Bruno was not executed because of his belief in heliocentricity but nonetheless, it had a real chilling effect.

Galileo Galilei (1564–1642) endorsed heliocentrism in his *Dialogue Concerning the Two Chief World Systems* (1632). Cardinal Roberto Bellarmine enjoined Galileo to neither "hold nor defend" heliocentrism in 1616. The latter was ordered to appear before the Holy Office in Rome in 1632. He argued that heliocentrism was not contrary to Scriptures. Galileo stood trial on suspicion of heresy in 1633. He was found guilty; his sentence demanded that he recant his heliocentric ideas; this he did. He was ordered imprisoned; this sentence was commuted to house arrest. His *Dialogue* was added to the *Index*, but was removed in 1835. He is said to have muttered upon hearing his sentence: *"Eppur si muove"* (and yet it moves).

Francis Bacon wrote in *The Essays: Of Atheism* (1625): "... a little philosophy inclineth man's mind to atheism; but depth in philosophy bringeth men's minds about to religion." René Descartes' *Discourse on the Method* (1637) was placed on the *Index* in 1663. Robert Boyle (1627–1691) was director of the East India Company and promoted Christianity in the East. He established the Boyle lectures to defend Christianity against "notorious infidels, namely atheists, deists, pagans, Jews and Muslims." Antony van Leeuwenhoek (1632–1723) was a Dutch Calvinist. He regarded his early microscopic studies of capillaries as proof of the great wonder of God's creation. Jan Swammerdam (1637–1680) argued that studying the Earth's creatures revealed the greatness of God. Isaac Newton (1642–1727) conceded that: "Gravity explains the motions of the planets, but it cannot explain who set the planets in motion. God governs all things and knows all that is or can be done."

Thomas Aikenhead (~1678–1697), a young student at the University of Edinburgh, was charged with blasphemy. "That ... the prisoner had repeatedly maintained, in conversation, that theology was a rhapsody of ill-invented nonsense ..." "That he rejected the mystery of the Trinity as unworthy of refutation; and scoffed at the incarnation of Christ." Aikenhead pleaded in vain for mercy on the gallows.

Carl Linnaeus (1707–1778) responded to the charge of impiety by the Archbishop of Uppsala: "It is not pleasing that I place Man among the primates, but man is intimately familiar with himself." "If I called man a simian or vice versa I would bring together all the theologians against me." Georges-Louis Leclerc, Comte de Buffon (1707–1788), denied Noah's flood. He was condemned by the Church and his books burned.

Joseph Priestly (1733–1804), one of the founders of Unitarianism, was branded an atheist in 1782. His *The Importance and Extent of Free Enquiry* (1785) roused a mob to burn his house and church. He fled England for the United States in 1791. The Priestley Medal, established in 1922, is the highest award of the American Chemical Society. Robert Schofield maintained that "Priestley was never a chemist; in a modern, and even a Lavoisian, sense, he was never a scientist. He was a natural philosopher, concerned with the economy of nature and obsessed with an idea of unity, in theology and in nature."

Louis Pasteur (1822–1895), near the end of his distinguished career, said: "The more I know, the more nearly is my faith that of the Breton peasant." Johann Gregor Mendel (1822–1884) performed his "Experiments on plant hybridization" in the Augustinian Abbey of St. Thomas in Brno, establishing the science of genetics (*Proc. Nat. Hist. Soc. of Brünn*, 1866). His subsequent duties as abbot left him no time to continue his science.

These varied examples do not permit a simple conclusion; however, several themes emerge. Most scientists, whether Catholic or Protestant, professed a true belief in God. Many felt that their science revealed the wonders of God's creations. The worst conflicts with the Church often involved personal disputes or heresies only peripherally related to science. The two real challenges to doctrine and beliefs were heliocentricity (Chapter B3) and evolution

(Chapter C14). The response of the Church to perceived dissenters became less violent with the passage of time.

Karl Popper (Chapter A10) emphasized the "... distinction between a scientific revolution and the ideological revolution which may sometimes be linked with it." He referred to Copernicus and Darwin: "... in these two cases a scientific revolution gave rise to an ideological revolution [...] ideological insofar as they both changed man's view of his place in the universe." The four basic equations of electricity and magnetism (Chapter B8) as summarized by Maxwell (1831–1879) were of corresponding scientific importance; however, they were not "ideological."

A10

Falsifiability: Karl Popper

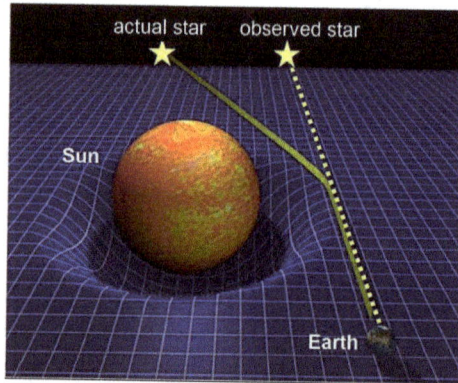

The deflection of starlight by the mass of the Sun.

Those who favor an operational definition, or characterization, of science often turn to *The Open Society and Its Enemies,* by Karl Popper. In essence he argued, as had several before him without full elaboration, that a theory in science can never be fully verified or proven. Its validity rests on the number and stringency of attempts to falsify it. In practice, a "control" experiment can be described as an attempt at falsification.

Sorry, there is no clear definition of or distinction between generalizations, patterns, hypotheses, theories, and laws. The validity or accuracy of these terms often depends, implicitly or explicitly, both on the range of phenomena discussed and on the attempts at falsification.

Of greater concern is how one applies falsification to historical sciences, such as (aspects of) astronomy, geology, and especially biology. Usually one cannot repeat the experiment under different conditions — the standard procedure in falsification (or verification). Given the spectrum of observations and experiments on components of the system, one seeks the most parsimonious interpretation, i.e. with the fewest and least improbable assumptions — one statement of Occam's razor (*lex parsimoniae*).

Popper did not explicitly address engineering and medicine. Even so, it is certainly good practice to make test runs or clinical trials before going to market.

Rarely does one elaborate on a new idea or make a discovery without precedent. The discovery of X-rays in 1895 by Wilhelm Röntgen (1845–1923) is a lovely exception. Inherent in most theories are implicit or explicit predictions. As critics of Popper, for example Thomas Kuhn (Chapter A11), argue, the failure of one of these predictions is seldom so decisive as to falsify the theory. Certainly a great deal depends on context and on whether a modification of the theory might ensure its survival.

Pascal's brother-in-law, Florin Périer, took a mercury barometer designed by Evangelista Torricelli (1608–1647) to the top of Puy de Dome, 1,000 m high, and left its identical mate at a lower level. He noted the difference "in barometric pressure of about seven cm; air had mass as predicted." Had no difference been observed, the concept of air having mass should have been accepted as falsified by the community.

Darwin himself set the standard when he acknowledged: "If it could be demonstrated that any complex organ existed which could not possibly have been formed by numerous, successive, slight modifications, my theory would absolutely break down." (*Origin of Species*, p. 171)

Popper frequently referred to the idea that "All life is problem solving," and explored how one goes about characterizing and solving problems in science, then in society. He concluded by arguing that an "open society" is better equipped to evaluate, or attempt to falsify, policies than is a rigid, authoritarian society. He did not argue that the policies of an open society are better than those of an authoritarian one. Setting and evaluating policy is an ongoing and never-ending process. In an open society one can more easily acknowledge errors and make corrections.

Popper continued his criticism of authority: "Knowledge in this objective sense is totally independent of anybody's claim to know; it is also independent of anybody's belief, or disposition to assert, or to act. Knowledge in the objective sense is knowledge without a knower; it is knowledge without a knowing subject." The authority of the knower is irrelevant (Notturno, *Science and the Open Society: The Future of Karl Popper's Philosophy*, 2000).

He was critical of induction as a valid way of doing science, without acknowledging the distinction made by Francis Bacon, who offered the analogy that ants merely gather, a simple accumulation of observations. However, bees not only gather but also use their own creative energies to synthesize honey.

Popper asserted that "Scientific knowledge need not be justified." Having dismissed induction, as well as *a priori* knowledge, he explored "critical rationalism" and concluded that falsifiability is its core. "Science appeals to experience to criticize its theories and not to justify them." He quoted Hume:

> "The only thing that the validity of an argument tells us regarding the truth of its premises and conclusions is that it cannot be the case that the premises are true and the conclusions are false."

Popper heard Einstein's lecture in 1919 in which he explained "How theory (relativity) might be tested." The *bending* of starlight passing near a massive sun would cause the star's apparent position to a viewer on Earth to be shifted by a calculable amount. During the solar eclipse on 29 May 1919, Arthur Eddington (1882–1944) sent expeditions to the island of Príncipe, off the west coast of Africa, and to Sobral, in Brazil, to observe stars in the Hyades cluster. The predicted deviation of about 1.0 arc seconds was observed (Eddington and Dyson, 2009). This test, i.e. attempt to falsify, inspired Popper to elaborate on the idea: "Scientists cannot discover and justify their theories through observation. But they can invent their theories as speculative solutions to their problems and then test them against observation and experience."

He argued for an ongoing process of proposals, leading to theories, leading to criticism (attempts at falsification). This would lead to new proposals and to new theories. He did not elaborate on the creative process whereby the experiments and/or observations might lead to new proposals.

Popper was a professor at the London School of Economics from 1949–1969. Perhaps his most distinguished student was George Soros, the extremely successful and wealthy investor. Soros questioned the application of the term, *science*, to social studies such as economics, because the objects of study, human beings, are themselves aware of such studies and could modify their behaviors accordingly.

A11

Paradigm: Thomas Kuhn

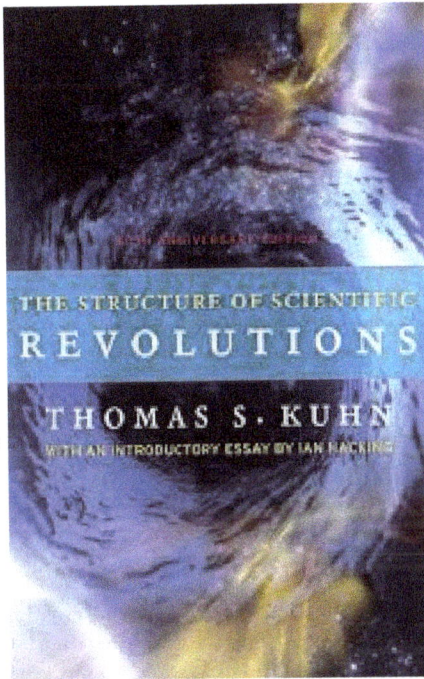

Thomas Samuel Kuhn
(1922–1996).
Shift happens.

Thomas Kuhn, in *The Structure of Scientific Revolutions* (1962), argued that from a historical perspective, fields of science can be viewed as functioning within a generally accepted, conceptual and technical framework — a paradigm — punctuated by abrupt shifts. He cited examples — Copernican heliocentricity, dynamics of Galileo and Newton, the overthrow of phlogiston by Lavoisier, Darwinian evolution, and Einstein's relativity — that are consistent with his formulation. These generally acknowledged revolutions were preceded by an accumulation of observations and/or experiments that were inconsistent with the prevailing paradigm. They were made by younger men not yet professionally invested in the old paradigm. Two of these examples, heliocentricity and dynamics, were major components of the scientific revolution of 1600–1700. Further, Kuhn argued that a single inconsistency or falsification did not lead to the rejection of the old theory as posited by Popper.

Several concerns have been expressed. Did some discoveries or theories, e.g. radioactivity and nuclear decay, replace old ones, i.e. was there a paradigm shift, or was there no pre-existing paradigm? If not, would that imply "pre-science" in that nascent discipline? The theory of plate tectonics, as the mechanism underlying continental drift, was quickly accepted by both young and old soon after "Magnetism in the Sea-floor" was published by Frederick Vine in 1963. Advances in molecular and cellular biology since 1950 have been dramatic. Did they involve the accumulation of new data, "normal science," at an accelerated pace or did they reflect a paradigm shift(s)?

Kuhn had to defend himself from critics who argued that he had acknowledged that the choice of paradigms was influenced by factors other than objective evaluation of available data and hence scientists held no privileged position of rationality relative to humanists.

Kuhn wrote that "… normal scientific research is directed to the articulation of those phenomena and theories that the paradigm already supplies." It is "puzzle-solving" with one of three goals:

1) determination of significant fact; 2) matching of fact with theory; 3) articulation of theory. Few scientists would disagree.

Thomas Kuhn (1922–1996) completed his Ph.D. in physics at Harvard in 1949. He was encouraged by James Conant, president of Harvard, to pursue the history of science. *The Structure of Scientific Revolutions* was published in 1962 and generated controversies that are still relevant a half century later. A summary of his arguments puts its antecedents and reverberations in perspective.

Much of Kuhn's thesis concerned the characterization of a "paradigm" and in turn, a paradigm shift. He did not give a concise, explicit definition but elaborated on many characteristics. He referred to "pre-science" in which there is no paradigm. A paradigm, characteristic of normal science, within any (sub)discipline consists of the:

- puzzles that the community tries to solve
- concepts and theories that practitioners consider important
- techniques and procedures commonly employed
- ways in which results are summarized and presented
- patterns of interactions among members of that community.

Just as the characteristics of a paradigm are a bit fluid, so the range within any discipline varies. "… the existence of a paradigm need not even imply that any full set of rules exists." These characteristics are not explicitly stated; as young scientists enter the field they pick them up.

Kuhn then elaborated on circumstances that presage a paradigm shift. He did not address the origin(s) of the original paradigm, the inferred essential step proceeding from pre-science to science. The community may encounter situations both in science and in infrastructure that are not accommodated by the existing paradigm. Herbert Butterfield in 1949 (*Origins of Modern Science*) had already described the challenge "… where one cannot escape an anomaly, and the theory has to be tucked and folded, pushed and pinched, in order to make it conform with the observed facts." Kuhn continued: "To be accepted as a paradigm, a theory must seem better than its competitors, but it need not, and in fact never does, explain all the facts with which it can be confronted."

Most scientists are not offended by Kuhn's statement that "Mopping-up operations are what engage most scientists throughout their careers." This mopping-up is equated with "puzzle solving." "It is no criterion of goodness in a puzzle that its outcome be intrinsically interesting or important." A "... striking feature of the normal research problems ... is how little they aim to produce major novelties, concepts or phenomena." "Even the project whose goal is paradigm articulation does not aim at the *unexpected* novelty." This would be a tall order for any scientist.

These unanticipated, and unanticipatable, novelties lead to crisis within the community. "The significance of crises is the indication they provide that an occasion for retooling has arrived."

He continued: "... a scientific theory is declared invalid only if an alternate candidate is available to take its place." "Nevertheless, anomalous experiences may not be identified with falsifying ones. Indeed, <u>I doubt that the latter exist</u>." "No process yet disclosed by the historical study of scientific development at all resembles the methodological stereotype of falsification by direct comparison with nature." "They (scientists) will devise numerous articulations and *ad hoc* modifications of their theory in order to eliminate any apparent conflict."

"Once a first paradigm through which to view nature has been found, there is no such thing as research in the absence of any paradigm. To reject one paradigm without simultaneously substituting another is to reject science itself." "... paradigm-testing occurs only after the persistent failure to solve a noteworthy puzzle has given rise to crises."

Kuhn wrote that these crises "close in one of three ways": i) "... normal science ultimately proves able to handle the crisis ..."; ii) "The problem is labeled and set aside for future generations ..."; iii) "... the emergence of a new candidate for paradigm ..."

"Almost always the men who achieve these fundamental inventions of a new paradigm have been either very young or very new to the field whose paradigm they change." (Contemporary feminists might subject him to a different sort or paradigm shift.)

Kuhn argued that in order to do science the community must have a paradigm — a shared language, common goals, and agreed problems. No matter that the paradigm is subsequently replaced; the information, protocols, and techniques would prove valuable working in the new paradigm. He cited Francis Bacon: "Truth emerges more readily from error than from confusion."

As did Popper, Kuhn extended his model beyond science. "In both political and scientific development the sense of malfunction that can lead to crisis is prerequisite to revolution." "Political revolutions aim to change political institutions in ways that those institutions themselves prohibit."

The following statements generated the greatest interest outside of the research community. "Like the choice between competing political institutions, that between competing paradigms proves to be a choice between incompatible modes of community life. Because it has that character, the choice is not and cannot be determined merely by the evaluative procedures characteristic of normal science ..." "... like the issue of competing standards, that question of values can be answered only in terms of criteria that lie outside of normal science altogether, and it is that recourse to external criteria that most obviously makes paradigm debates revolutionary." Scholars outside of the research community cited this to argue that scientists were no more analytical or rational and that they should relinquish their sense of privilege. Kuhn defended himself: "My remarks have been misconstrued." "In debates over theory-choice; once premises have been accepted — the only analysis is one of logic. In contrast discussion of the premises inevitably introduces subjective value judgments. However, this does *not* mean that logic is not a factor."

Imre Lakatos wrote in *Falsificationism and the Methodology of Scientific Research Programs* (1962): "The clash between Popper and Kuhn is not about a mere technical point in epistemology. It concerns our central intellectual values, and has implications not only for theoretical physics but also for the underdeveloped social sciences and even moral and political philosophy."

In contrast, Paul Feyerabend in *How to Defend Society against Science* (1978) was dismissive: "[Kuhn's] ideology of science could only give comfort to the most narrow-minded and the most conceited kind of specialism. It would tend to inhibit the advancement of knowledge. And it is bound to increase the anti-humanitarian tendencies which are such a disquieting feature of much of post-Newtonian science."

The term *paradigm shift* is now so widely applied as to have lost its original meaning. Yet, historians still ask whether a change or an advance in a (sub)discipline of science reflects a shift of paradigm or accumulation of insight and technique within *ordinary science*. Is this a meaningful question?

The most important concern from the perspective of this book is whether a synthesis of Popperian and Kuhnian perspectives informs contemporary science, and especially biology. Both implicitly address "pure" science. One is left to wonder whether their analyses and arguments apply to engineering and medicine.

A12

Two Cultures: C.P. Snow

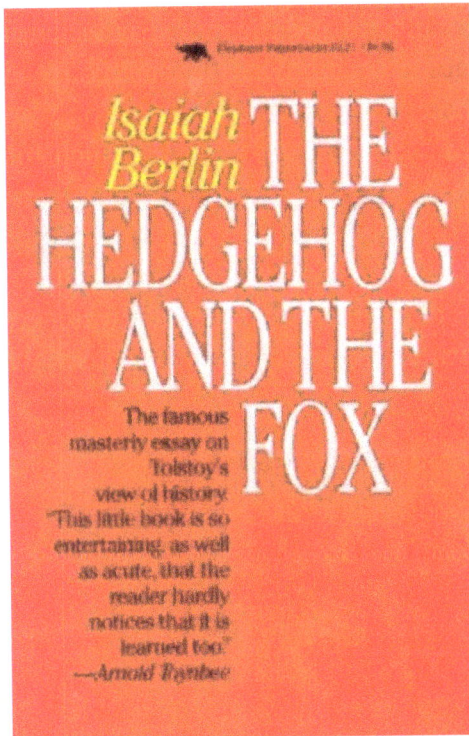

After Archilochus "the **fox** knows many things, but the **hedgehog** knows one big thing."

C.P. Snow, British physicist turned novelist, delivered the Rede Lecture in 1959. Much of his lecture addressed education in Britain and compared it with the upstart Russian and American systems. He also said that "... the intellectual life of the whole of Western society is increasingly being split into two polar groups," scientists versus humanists, and that the two were separated by a "gulf of mutual incomprehension." "The non-scientists have a rooted impression that the scientists are shallowly optimistic, unaware of man's condition." Snow subsequently acknowledged that his introductory remarks contrasting the intellectual traditions and mutual incomprehension of the humanities and of the sciences were a great simplification.

He wrote in *The Two Cultures: A Second Look* (1964): "In fact, those two revolutions, the agricultural and the industrial-scientific, are the only qualitative changes in social living that men have ever known." He further argued that: "Industrialization is the only hope for the poor," and elaborated "… if you go without much food, see most of your children die in infancy, despise the comforts of literacy, accept twenty years off your own life, then I respect you for the strength of your aesthetic revulsion [of industrialization]." One can argue that his defense of science and its impact was a gross simplification, yet anyone yearning for a simpler, pre-industrial life must address his arguments.

The phrase "two cultures" was catchy and the ground fertile. For the past half-century numerous scholars have addressed implied binary divisions within society, or within the academy, or even within a discipline, e.g. history vs. political science. The idea was extended to "culture wars," then to all sorts of scholarly "wars." If said scholars had actually experienced the horrors of war, they might not have used the term so glibly.

Snow extended this binary concept: "Pure scientists and engineers often totally misunderstand each other." He acknowledged that many others, for example Jacob Bronowski in *Science and Human Value* (1956), had already addressed these issues. Snow argued: "But I believe the pale of total incomprehension of science radiates its influence on all the rest."

Snow did, however, capture a kernel of truth. During the Renaissance the very small fraction of the population who were literate probably considered natural history to be an integral part of their intellectual world. The encyclopédistes of 18th century France felt that at least a few individuals could grasp all of human knowledge. This knowledge continues to expand; few scientists today are familiar with the full breadth of science, few humanists with all of the arts. Does the concept of "two cultures" simply reflect the breadth of contemporary knowledge and our relatively few, $\sim 10^{11}$, neurons? Or is there, on average, a real and significant difference between what scientists value and the way that they think in comparison with the rest of the population? At higher resolution, is there a difference between biologists and physicists, or between those focused on application (engineers and physicians) and those seeking basic knowledge?

Snow's rather simple formulation basically asked how the humanist and the scientist might better understand and value the other's work, that is, how to encourage and appreciate the dialog. Other scholars, most notably E.O. Wilson in *Consilience:*

The Unity of Knowledge (1998) have addressed a fundamentally different but related question. Is the nature of knowledge and understanding in the humanities and in the sciences inherently different; will it be in the future? Might future advances in molecular and neurobiology actually permit one to make meaningful comments about the humanities based on science? Would (aspects of) the humanities be subsumed under the broader umbrella of science?

It is often easier to develop an argument assuming a bimodal distribution — blue vs. red; left vs. right — when in fact there is no cleavage or the distribution is tri-modal. The legend of the Hedgehog and the Fox, attributed to Archilochus, says "... the fox knows many things, but the hedgehog knows one big thing ..." Isaiah Berlin in 1953 examined the single-minded focus of Dosteovsky versus the range of interests shown by Tolstoy. Granted, this division is a gross simplification; however, it does facilitate a description of different perspectives. Few people or institutions are purely blue or red, but some complex mixture and synthesis — shades of purple.

The above-mentioned encyclopédistes were a group of 18th century writers in France who compiled and wrote the *Encyclopédie*, edited by Denis Diderot and Jean le Rond d'Alembert. Many were part of the intellectual group known as the philosophes. They promoted the advancement of science and secular thought and supported tolerance, rationality, and the open-mindedness of the Enlightenment. One would hardly apply the epitaph "two cultures" to them or to their encyclopedia. It is difficult to accept that this perspective is gone forever, since it is the theme of this book.

A13

Emergence

Mandelbrot set.

Several inter-related terms or concepts are essential to discussions of science. There is not complete agreement about their definitions and implications; the following reflect something of a consensus.

Reductionism is associated with simplification. Structures or properties at one level are described or understood by characteristics of a more basic level. This inter-level approach has been equated with fundamental understanding in physics.

Complexity refers to a system or process that can be adequately described only by many degrees of freedom and a broad range of values of the associated parameters.

Predictability refers to the ability to predict the characteristics of a system at some future time, given any or all of the characteristics of that system at an earlier time.

Determinism refers to a system or a process whose future state can, in principle, be predicted from knowledge of its state at an earlier time.

Chaos refers to a system that is deterministic but unpredictable in practice due to its extreme dependence on initial conditions.

Quantum Uncertainty derives from the "Uncertainty Principle" of Heisenberg, $\Delta x \cdot \Delta p$ (or $\Delta E \cdot \Delta t$) $> \hbar/2$.

Randomness: the outcome of a single coin toss is probabilistic, or random; the distribution of many tosses is highly predictable.

Emergence refers to a property of a system that is assumed to have arisen from more fundamental characteristics but cannot be explicitly derived or predicted from those characteristics.

Vitalism in its various forms ascribes properties to living systems that are unique and not derivable from the properties of its non-living components.

Reductionism: Steven Weinberg (Nobel Laureate, 1979) was certainly right (almost): "All the explanatory arrows point downward, from societies to people to organs, to cells, to biochemistry, to chemistry, and ultimately to physics." For many this is the essence of, if not the definition of, science. As will be elaborated, most molecular and cellular biology can be cast in this reductionist framework. Evolutionary biology requires quite a different perspective. The wedding of these two views makes contemporary biology *especially* exciting!

Complexity refers to a system or process that can be adequately described only by many degrees of freedom and by a broad range of values of the associated parameters. The skill of a good scientist is reflected in his choice of a simple system amenable to investigation with available techniques. This pushes the more complex problems on to the next generation, hence Vannevar Bush's *Endless Frontier* (Chapter C16). Many complex systems in biology are now analyzed by statistical analysis of massive data sets. Systems that have emergent properties are always complex.

Predictability (or of greater concern, unpredictability) or *determinism* refers to the ability to predict the characteristics of a system, often but not necessarily *complex*, at some future time, given any or all characteristics of that system at an earlier time. Pierre-Simon Laplace (1749–1827) was keenly aware of the problem: "… imagine an Intelligence who would know at a given instant of time all forces acting in nature and the position of all things of which the world consists … Then it could derive a result that would embrace in one and the same formula the motion of the largest bodies in the universe and of the lightest atoms. Nothing would be uncertain for this Intelligence." (Chapter B4)

Determinism refers to a system or a process whose future state can in principle be predicted from knowledge of its characteristics at an earlier time. If a system is deterministic, one can in principle predict its future state.

Chaos refers to a (mathematical) system that is deterministic but unpredictable in practice due to its extreme dependence on initial conditions. Mathematicians have explored those points or regions in a multi-dimensional space of parameters that are (potentially) chaotic.

Quantum Uncertainty: One can predict with extremely high accuracy the spatial distribution of photons in a diffraction pattern; however, one can make only a probabilistic statement about the path chosen by a single photon. One can assign a very accurate half-life for decay to a collection of identical radioactive atoms; yet one cannot predict when a single atom will decay.

A related, fundamental constraint is imposed by the Heisenberg uncertainty principle: $\Delta x \cdot \Delta p$ (or $\Delta E \cdot \Delta t$) $> \hbar/2$ is a fundamental physical limitation. As one measures x to ever greater precision, the error, Δx, grows ever smaller; correspondingly, one loses information about p; Δp must grow larger. Planck constant, $h = 6.626 \times 10^{-27}$ erg ($\hbar = h/2\,\varLambda$); x = position; p = momentum; E = energy; t = time.

Randomness of a single coin toss or of a single roll of a dice (die for sticklers) is probabilistic; the distribution of many tosses or of many rolls is highly predictable. That distribution can be described by the binomial distribution (Pascal's triangle) — 1 1 1 1 2 1 1 3 3 1 1 4 6 4 1 — which in the limit is described by the normal distribution (Chapter B2).

Emergence is both a valid and a valuable concept as one addresses more complex systems. To approximation one might consider Weinberger's examples in reverse. Given all the information one might wish about atomic physics, one cannot predict all the structural and reaction characteristics of molecules. Given all the information one might wish about chemistry, one cannot predict all the structural and functional characteristics of proteins or nucleic acids. Given all the information one might wish about bio-macromolecules, one cannot predict all the characteristics of cells. These statements of our limitations continue up to organs, to people, and to societies (Kaufman, 2008). This characterization begs the question of whether an emergent property today will be predictable in the future. Is emergence merely a statement of contemporary ignorance? In practice one wisely uses the concept of maximum parsimony (one of many restatements of what is often referred to as Occam's razor). It is the explanation or interpretation of existing observations or results that involves the fewest (or smallest or most reasonable) assumptions. The most parsimonious interpretation(s) may change, within the existing paradigm (Chapter A11), as new data becomes available.

Vitalism was nominally refuted by Friedrich Wöhler, who synthesized urea, $CO(NH_2)_2$, from ammonium cyanate, $NH_4 NCO$, in 1828 (Chapter C8). He wrote to Berzelius about "The great tragedy of science, the slaying of a beautiful hypothesis by an ugly fact." No biologist should use the term *vitalism* before age 40 or until having received tenure. Nonetheless, many vital characteristics, such as tissue excitability, might be legitimately described as emergent properties. A single falsification, after Popper, did not in itself spell the doom of vitalism. Whether its demise conforms to Kuhn's formulation of a paradigm shift is problematic.

Francis Crick (Nobel Laureate, 1962) was unequivocal: "And so to those of you who may be vitalists I would make this prophecy: what everyone believed yesterday, and you believe today, only cranks will believe tomorrow." Roger Sperry (Nobel Laureate, 1981) was more sympathetic: "The events of inner experience, as emergent properties of brain processes, become themselves explanatory, causal constructs in their own right, interacting at their own level with their own laws and dynamics. The whole world of inner experience (the world of the humanities) long rejected by 20th century scientific materialism, thus becomes recognized and included within the domain of science." Whether humanists welcome this inclusion is another question.

Ernst Mayr (1904–2005) sought a balance:

> It would be ahistorical to ridicule vitalists. When one reads the writings of one of the leading vitalists like Driesch one is forced to agree with him that many of the basic problems of biology simply cannot be solved by a philosophy as that of Descartes, in which the organism is simply considered a machine ... The logic of the critique of the vitalists was impeccable. But all their efforts to find a scientific answer to all the so-called vitalistic phenomena were failures ... rejecting the philosophy of reductionism is not an attack on analysis. No complex system can be understood except through careful analysis. However, the interactions of the components must be considered as much as the properties of the isolated components.

Often physicists or ecologists, in the privacy of their own labs or plots, will refer to a device or to a plant as having a will or an urge. One should distinguish a manner of speaking from a deep philosophical commitment.

Section B

Physical Sciences: Overview

The main focus of this book is biology; however, the development of biology has been, and very certainly will continue to be, highly dependent on advances, both conceptual and technical, in the physical sciences. The emphasis of the chapters is this section is to understand how these disciplines of physics affected biology up to the time of Darwin and during the following decades.

The term, *physics*, is used here as an abbreviation for all of the physical sciences and mathematics as well as their applications, collectively called *engineering*. The emphasis will be on the early advances in these various disciplines, pre-1900, i.e. before relativity, quantum mechanics, and nuclear physics. How did physics impact biology and inform its development? Physics in the 20th century will be mentioned as it relates to a few selected examples from contemporary biology.

Most of physics adopts a reductionist perspective; that is, to understand and explain physical phenomena in terms of ever more fundamental particles and forces. However, geology and astronomy have a historical perspective; how does one interpret past events that one cannot replicate? Applications of physics, engineering, are inseparable from "pure" physics, that is, understanding Nature. If they are indeed inseparable, can one apply the same philosophical concepts to application as to understanding?

Many physicists would shudder at the suggestion of including devices in discussions of physics. Was the development of the area detector (Charpak, Nobel Laureate 1992), or the bubble chamber (Glaser, Nobel Laureate 1960), or a telescope with a 20x lens (Galileo, 1608) mere gadgetry? Did the engineering of these men have as great an impact on their contemporary societies as did the first secure attachment of stone to haft in the Neolithic?

 B1. Engineering
 B2. Mathematics
 B3. Astronomy
 B4. Mechanics
 B5. Alchemy
 B6. Phlogiston
 B7. Periodic Table
 B8. Electricity, Magnetism, and Optics
 B9. Thermodynamics
B10. Geology

The inclusion of engineering in physics has profound implications for both practice and administration. Francis Bacon in 1620 wrote of *works*; science should serve the needs of society (Chapter A6). King Charles II granted a royal charter to the Royal Society of London in 1640, anticipating benefits to the kingdom. Many would agree with C.P. Snow that even though there was a lot of negative impact, the industrial revolution of the 1700s ultimately benefitted Britain and the world (Chapter A12). That revolution required energy; much of it was used to drive steam engines. The optimization of those engines drove the development of thermodynamics, the concept of entropy (Chapter B9), and then information theory (Chapter C16).

This concern with the applications of science to engineering and medicine underlay the success of Vannevar Bush in establishing the precursors of the National Science Foundation and National Institutes of Health in the U.S. He coined the memorable phrase "Endless Frontier" well before Popper described the cycle of proposal, theory, criticism, (new) proposal, (new) theory... (Chapter A10). Grant proposals to NSF or to NIH request a statement of justification or application. This is not an unreasonable request for use of taxpayers' money. The real controversies concern the fraction of big winners needed to cover the majority of projects with no immediate application as well as the time scale — years, decades, or centuries — before the "works" of Francis Bacon are realized.

Many of the historical and philosophical concepts explored in Section A implicitly or explicitly addressed problems in pure physics. To what extent are these concepts applicable to engineering as well as to biology and medicine? Our ideologies were challenged by both a heliocentric model of the universe and by the concept of evolution by natural selection. The industrial revolution changed the ways in which many people in the West lived;

standards of behavior and governance struggled to keep pace. Today gene sequencing and manipulation are offering their own unique challenges.

Testing new equipment and structures as well as clinical trials for new drugs and prostheses is not only good practice but often mandated. These tests are attempts at falsification (Chapter A10). Although difficult to define, most subfields of engineering and medicine have their own paradigms; however, the boundaries may be a bit more fluid.

Given this broad view of engineering, one asks about its origins. The ability to make fire and to knap flint probably had a greater impact on that contemporary society than any feat of engineering since.

Some level of scientific acumen appears to be genetically encoded. Imagine making a nest of twigs that will withstand heavy winds and swaying branches, using only your beak. Sophisticated nest making is observed in insects and mollusca as well as throughout the vertebrates.

Many birds as well as mammals make and use tools. As reported by Auersperg *et al.* (2012):

> Figaro, cockatoo, dropped the nut through the aviary's wire mesh, where it landed on a wooden beam. He tried to retrieve the nut with his claw. After trying and failing to reach the nut with a tiny stick he found on the aviary floor, Figaro set about making what he needed. With his beak, he ripped off a large splinter from the wooden beam, a task that took him almost 25 minutes. Holding his tool in his beak, he eventually raked the nut to a spot where he could grab it with his tongue and beak.

Many mammals eat special foods to counter poisons or to reduce the impact of infection. When experiencing stomach upset, chimpanzees turn to noxious plants, such as the bitter leaf, to help eliminate any intestinal parasites that might be causing their symptoms. "They remove the bark and leaves like a banana," explained Michael Huffman of the University of Kyoto. "Once they suck out the juice, the chimps spit out the fiber" — receiving no nutritional benefit from the plant, but giving the nematodes in their guts a toxic bath. Science, broadly defined to include applications to engineering and medicine, is not a uniquely human pursuit.

We can ask the same question of art. One marvels at the sophistication of the drawings deep in the Lascaux cave made by our ancestors 17,000 years ago. Although some animals might be said to appreciate beauty, for instance in courtship displays, there is no evidence of any species, other than our own, doing art (Chapter D4). It is hardly a coincidence that the Renaissance preceded the scientific revolution.

B1

Engineering

Alice Auersperg, Birgit Szabo, Auguste von Bayern, and Alex Kacelnik, "Spontaneous innovation in tool manufacture and use in a Goffin's cockatoo." *Current Biology* 6th November (2012).

Making a timeline for engineering forces one to address several questions. Is the construction of any tool or tent engineering? Insects and birds build elaborate nests — are they engineers? Spiders and mollusks build webs and shells of their own substance. Birds and apes use thorns or straws as implements; these behaviors seem innate.

One might define engineering as practical application of insights from physics; as one describes medicine, and agriculture, as applications of biology. But does building a shelter or dressing a wound reflect an understanding of the natural world, of "pure" science? Can engineering behaviors be genetically encoded without necessitating or reflecting any real understanding of "basic" physics?

What are the criteria for the significance of technical innovation — impact on society or cleverness of invention? By whatever metrics, the judgments depend on context. Does such a timeline provide new insights or does it merely reflect pre-existing prejudices?

A few achievements can be assigned a specific time, place, and designer, for example, J.A. Roebling and the Brooklyn Bridge, completed in 1883, or the Wrights' flight at Kitty Hawk, 1903. To whom do we attribute design and construction of Pont du Gare by the Romans? Most feats actually involve processes spanning generations and often involving many, unrecognized inventors, e.g. the stonemasons of the Cathedral of Chatres or employees of the early Bell Labs. In contrast, major achievements of fundamental or "pure" science, and of art, can usually be attributed to an individual.

Neither Karl Popper nor Thomas Kuhn, nor most philosophers and historians of science, include engineering in their analyses. This book adopts the minority view that it is misleading to separate the lens grinder from the star gazer, especially if they are one and the same person.

Champions of science from Francis Bacon to Vannevar Bush have argued for the value of "works" to society. These works are often realized in the applications of engineering or medicine. This generalization begs several questions. How does one measure the benefits to society of these applications; how valid is the verbiage in the "Impacts" section of grant applications? To what extent and in what ways does this process impact the general intellectual climate of society? Or, more humbly, in what ways does this climate facilitate science? After all, the Renaissance preceded the scientific revolution. Leonardo perhaps contemplated art more deeply that he did levers.

The distinction between research in technical development and pure science is valid in a conceptual sense. Many discussions of history and/or philosophy of science are couched, explicitly or implicitly, in terms of "pure" science, i.e. understanding the natural world. To what extent is it appropriate to extend these ideas to engineering, i.e. making use of the natural world?

Given all of these and other provisos, an overview of major achievements of engineering does suggest several generalizations:

1. Feats of engineering occurred, quite independently, in many different lands. Even so, the distribution of these achievements is hardly random; some places and some epochs were especially productive. There is a correlation with the general prosperity and vitality of a society and its inventions; assigning causality is more challenging.
2. Some inventions, but definitely not all, were dependent on insights provided by basic science; more so in recent centuries and with increased complexity and sophistication.
3. Often the challenge to the engineer drove the underlying science. Often the technical advances provided new questions or tools for science.

How to analyze or rank the more recent achievements of engineering? From the perspective of daily life, perhaps the flush toilet tops the list. If we're considering the impact of the discovery on pure science, perhaps grinding the first lenses by Robert Hooke (1635–1703) and by Antonie van Leeuwenhoek (1632–1723). Or the development of the transistor in the Bell labs by Bardeen, Brattain, and Shockley (Nobel Prize, 1956) in 1949 (*Crystal Fire*, 1998). Or the accompanying publication on information theory and the distinction between "information" (or entropy) and "meaning" by Claude Shannon (1916–2001), "Communication Theory of Secrecy Systems" in the *Bell System Technical Journal*.

From the perspective of the application of science, perhaps Tesla's alternating current power transmission and the electric motor? With these questions in mind, it is rewarding to consider some great feats of engineering.

The Roman Empire made few contributions to pure science; however, their contributions to civil, or military, engineering were monumental, e.g. *De architectura*, 10 volumes by Vitruvius (20 BC). The *Book of Knowledge of Ingenious Mechanical Devices*, by Al-Jazari (1206), describes many contributions from the world of Islam. *De re metallica*, by Georgius Agricola (Georg Bauer), translated to English by Lou Henry Hoover and Herbert Hoover, summarized the status of mining and minerals in 1556. The Royal Society of London, established in 1642, devoted great attention to "works" as described by Francis Bacon. The early development of the steam engine to pump water from mines (1712), by Thomas Newcomen, and subsequently to drive locomotives (1769), by James Watt, gave Britain a head start in the industrial revolution.One can hardly assign causality but nonetheless, the sequence is suggestive: Renaissance, 1500–1600; scientific revolution, 1600–1700; industrial revolution, 1700–1800.

The cotton gin, invented by Eli Whitney in 1793, made slavery in the United States profitable, far more so than had sugar cane in the Caribbean. The first petroleum refinery near Pittsburgh, established by Charles Lockhart, for crude skimmed from salt wells in 1852, produced over four barrels a day of various distillates, primarily lamp oil. The refinery was ready for the first oil well near Titusville, PA, in 1859, operated by Edwin Drake.

The military has contributed in its way, to both construction and destruction. Building fortifications, roads, and bridges to serve the military was the main task of civil engineering; Mr. Gatling's gun in 1862 might be considered an unwieldy assault weapon. Dynamite, as developed by Alfred Nobel in 1866, found many applications, constructive and destructive, and made him extremely wealthy.

1869 was a vintage year. The Suez Canal opened and the golden spike was driven at Promontory Point, Utah. The canal might have been completed two millennia earlier had the Egyptians so focused their considerable engineering energies. The spike absolutely depended on relatively recent advances in steam engines and steel production. Which had a greater impact on society?

1888 could compete. A radio wave receptor was built by Heinrich Hertz in Germany. Nicola Tesla built an induction, alternating current motor in New York and, much to the irritation of Mr. Bell, championed the transmission of AC, as opposed to DC, electricity.

Perhaps the Model T was not the best car produced in 1908, but its impact on society was far greater, in part because Henry Ford introduced mass production on moving assembly lines with completely interchangeable parts. The first television broadcasts in Germany and the U.S. in 1928 were not reckoned to contribute to the contemporary obesity epidemic — a great example of unintended consequences. One still marvels at the Hoover (Boulder) Dam, completed in 1936, on the Colorado River, dividing Nevada and Arizona.

As argued in the introduction to this section, application is an integral part of physics. Did the applied science of Charpak, Glaser, and Galileo have as great an impact on their contemporary societies as did the first secure attachment of stone to haft in the Neolithic? Did mathematics provide the link to analysis and physics? Many species of mammals and birds are natural toolmakers and users; given this broad perspective, engineering is hardly unique to humans.

B2

Mathematics

Inverted catenary.

Some of the basics of mathematics — counting, addition, subtraction, geometry — were employed independently by several civilizations. This assertion immediately raises still unresolved concerns. Is math created by humans or is it "out there in the natural world" to be discovered? Is doing basic math a uniquely human attribute? Some animals can count and communicate; how do they distinguish one from two from many objects? To what extent and how is this ability genetically encoded? Hunter-gatherers have words for "one, two, three, four ..." then on to "many". Who extended the concept of counting objects to numbers without a physical attribute?

Civilizations in Egypt, Mesopotamia, and China had mastered basic arithmetic and geometry, including the concept of a right angle and some approximation of trigonometric functions, before the Greeks. What seems to have characterized the Greeks was a commitment to abstraction for the pleasure of thinking beyond application. This sort of work avoidance is still practiced in poorly supervised universities. Plato maintained that: "... the world was God's epistle written to mankind..." and "... it was written in mathematical letters." How then to deal with zero and with infinity?

It was Greek mathematics, sometimes translated to Arabic, then to Latin, that provided a foundation for the Renaissance and scientific revolution in Western Europe. In what ways did concepts of mathematics encourage advances in physics and engineering? Conversely, how did problems in physics motivate mathematicians? Were the appropriate concepts of mathematics already available to biologists as they encountered new problems? Did the general climate of physics in general and mathematics in particular set a tone for biology?

Usually basic ideas in mathematics are encapsulated in terms of an equation. Robert Recorde, in *The Whetstone of Witte* (1557), first proposed the use of the symbol "=" both to assign a value, e.g. $x = 3$, and to show a functional relationship, $y = 2x + 1$. The origins of the concept of such an explicit equivalence, as opposed to a verbal statement, remain obscure. Some of these relationships, or equations, were purely mathematical and did not necessarily describe actual objects, processes or properties.

Ian Stewart nicely summarized his choices in *In Pursuit of the Unknown: 17 Equations that Changed the World* (2012). The following equations do not have explicit physical parameters.

$a^2 + b^2 = c^2$ (Kaplan and Kaplan, *Hidden Harmonies: The Lives and Times of the Pythagorean Theorem*, 2012).

$\log x \cdot y = \log x + \log y$. After John Napier (1550–1617) of Edinburgh; the basis of the slide rule.

$df/dt = \lim_{h \to 0} [f(t+h) - f(t)]/h$. "$t$" is usually time but could be any parameter. Leibniz and Newton have been squabbling for over three centuries about priority and notation.

$i^2 = -1$. Defines complex numbers and the complex plane.

$$\sin z = z - z^3/3! + z^5/5! - z^7/7! (z \text{ in radians})$$

$$\cos z = 1 - z^2/2! + z^4/4! - z^6/6!$$

$$e^z = 1 + z + z^2/2! + z^3/3!$$

$$e^{iz} = \cos z + i \sin z$$

$$e^{i/t} + 1 = 0.$$

Euler's identity, published in 1828.

Richard Feyman described it as "the most remarkable formula in mathematics."

$F_{aces} - E_{dges} + V_{ertices} = 2$. This characterization of a three-dimensional solid is the basis of topology.

$$\emptyset(x) = (2\pi\sigma)^{-0.5} \exp -[(x - \mu)^2/2\sigma^2]$$
$$\emptyset(x) = \text{probability of getting } x;$$
$$\sigma = \text{standard deviation}; \mu = \text{mean value}$$

This "normal distribution" is the limiting case of the binomial distribution, de Moivre (1733). e.g. $(p+q)^4 = p^4 + 4p^3q + 6p^2q^2 + 4pq^3 + q^4$, sometimes referred to as "Pascal's triangle." (1 11 121 1331 14641 ...) and first noted by Chandas Shastra in 300 B.C.

The wave equation, $d^2u/dt^2 = c^2 d^2u/dx^2$ (u = displacement; c = velocity of propagation; x = 3D space) describes, among many examples, the vibration of a taut string. As a special case, one can derive the relationship of musical whole tones in terms of vibrational frequencies (Chapter D5).

1	9/8	81/64	$(9/8)^3$	$(9/8)^4$	$(9/8)^5$	$2[\sim (9/8)^6]$
C	D	E	F	Ab	Bb	C

The Fourier (Joseph, 1768–1830) transform and inverse transform:

$$F(s) = \int f(\mathfrak{r}) \exp(i \cdot 2 \cdot \pi \cdot r \cdot S) dr$$
$$f(\mathfrak{r}) = t \int F(s) \exp -(i \cdot 2 \cdot \pi \cdot r \cdot S) ds$$

can describe any continuous function as the sum of sine and cosine waves.

These "pure" equations soon found application in the real world. It seems that mathematicians are often plagued by pesky scientists, wanting to use their concepts and equations.

Newton's three laws, summarized in Chapter B4, laid the foundation for the fields of mechanics and dynamics. They can be regarded as elaborations on the conservation of momentum. Combined with the postulate that the gravitational force of attraction

n is proportional to the two masses and to the inverse square of the distance between them: $F = g \cdot m_1 \cdot m_2 \cdot d^{-2}$ ($g =$ gravitational constant). He could formulate equations that describe the stable elliptical orbits of planetary motion described by Kepler. However, Newton conceded that he could not explain the origins of those orbits.

The Navier–Stokes equation of fluid motion is an extension of Newtonian physics:

$$\rho(dV/dt + V \cdot \nabla V) = -\nabla p + \nabla \cdot T + f$$

$\rho =$ density of fluid; $V =$ velocity of local flow; $p =$ pressure; $T =$ stress; $f =$ body forces.

Maxwell's equations (Chapter B8) summarized a century of research in electricity, magnetism, and optics: (James, 1831–1879)

$$\nabla \cdot E = 0 \quad \nabla \times E = -c^{-1} \, dH/dt$$

$E =$ electric; $H =$ magnetic field; $\nabla \cdot =$ divergence; $\nabla \times =$ curl

$$\nabla \cdot H = 0 \quad \nabla \times H = c^{-1} \, dE/dt$$

The second law of thermodynamics (things can only get worse), $dS \geq 0$ and the definition of entropy in term of microstates available to the system:

$$S = k \cdot \ln \cdot W, \quad k = 1.381 \times 10^{-23} \, \text{J} \cdot \text{K}^{-1}$$
$$[k, \text{Ludwig Boltzman}(1844–1906) \text{ constant}]$$
$$W = N!/\Lambda_i \, N_i!(\text{wahrscheinlichkeit frequency of,}$$
$$\text{macrostate, } W)$$
$$(df/dt)_{\text{force}} + (f \, df/dx)_{\text{diffusion}} + (F/m \, df/dv)_{\text{collision}}$$
$$= (df/dt)_{\text{total}}(\text{Boltzmann equation})$$

where $f =$ distribution function, single particle position and momentum at a given time; $F =$ force; $m =$ mass of the particle; $t =$ time; $v =$ average velocity of the particles.

Certainly no one in 1885 anticipated that a parallel formulation by Claude Shannon in the Bell labs in 1949 would equate "information" with entropy: information $= -_{i=1} \sum^n P_i \cdot \ln_2 P_i$ for n states and the probability, P, of the system being in that state (Chapter C16).

Einstein's $E = m \cdot c^2$ ($E =$ energy; $m =$ mass, $c =$ speed of light), 1905, was a component of his theories of special and general relativity (Albert, 1879–1955).

Schrödinger's equation, $i\hbar\, \partial\psi/\partial t = \hat{H} \cdot \psi$ ($\psi =$ quantum wave function; $\hat{H} =$ Hamiltonian operator; $\hbar = h/2 \cdot \pi; h = 6.62610^{-34}$ Joule sec) laid the foundation for understanding the interaction of electrons and protons, among other things, and modern chemistry (Erwin, 1897–1961).

The recursion formula describes population growth and can be applied to many processes: $x_{t+1} = k\, x_t\, (1 - x_t)$, where $x =$ population size; $t =$ present time; $t + 1 =$ next generation; $k =$ growth rate. The earliest example is the series of Fibonacci, nicknamed Leonardo Pisano Bigollo, (1170–1250):

$$F_n = F_{n-1} + F_{n-2}.$$

From biology to economics, the Black–Scholes equation describes, under "ideal free market conditions" the price of a commodity, S, in terms of the price of a financial derivative, V:

$$1/2\sigma^2 S^2 d^2 V/dS^2 + r \cdot S dV/dS + dV/dt$$
$$-r \cdot V = 0; \quad t = \text{time}; \quad \sigma = \text{volatility};$$
$$r = \text{risk free interest rate}.$$

These 17 equations capture some of the history of mathematics. Pythagoras, $\sim$530 B.C., realized that $2^{1/2}$ is irrational. Ptolemy of Alexandria wrote the *Almagest*, "the greatest" (book of science) about 150 A.D. Mathematicians in India $\sim$500 B.C. established the base 10 numeral system and a few centuries later introduced the concept of zero. Mathematicians in Mesopotamia, by $\sim$500 B.C., established the base 60 number system, e.g. 360° in a full circle.

Napier, $\sim$1615, defined logarithms and 'e' $\sim$2.718... the numerical base of the natural log system in *Mirifici Logarithmorum*

Canonis Descriptio. Rene Descartes, ~1619, introduced analytical geometry and the *Cartesian* coordinate system.

Isaac Newton, ~1665, and Leibniz, ~1673, introduced infinitesimal calculus. Newton's "fluxions" were cumbersome; we use Leibniz's notation today. Leibniz in 1680 used symbolic logic and in 1691 described ordinary differential equations. Christian Goldbach presented his conjecture, which remains unproven (every even number, >2, is the sum of two primes), in a letter to Euler in 1742. Adrien-Marie Legendre in 1805 introduced the method of least squares for fitting a curve to a given set of observations. George Boole, 1847, formalized symbolic logic in *The Mathematical Analysis of Logic*, defining what is now called Boolean algebra. There is no record in the National Security Administration of these guys having exchanged emails, but they were certainly up to no good.

The progress of computation and of the underlying mathematics is reflected in the *Life of Pi*:

2000	Mesopotamia	~3.125
1650	Ahmes	~3.16
261 B.C.	Archimedes	$3 + 1/7(\sim 3.1429)$ to $3 + 10/71(\sim 3.1408)$
450 A.D.	Zu Chongzhi	to 7 decimal places
1400	Madhava	to 11 decimal places
1424	Ghiyath al-Kashi	to 16 decimal places
1596	Ludolf van Ceulen	to 20 decimal places
1706	John Machin	to 100 decimal places
1735	de Moivre	derived an infinite series for π
1789	Jurij Vega	to 140 decimal places
1949	John von Neumann	2,037 places using ENIAC
1961	Shanks and Wrench	100,000 using IBM-7090
1987	Kanada *et al.*	134,000,000 using NEC SX-2
2002	Kanada *et al.*	124,100,000,000 using Hitachi 64-node

This is just one of scores of infinite series (and infinite products) expressing the numerical value of π.

$$\Pi = \left(6\left[1^{-2} + 2^{-2} + 3^{-2} + 4^{-2}\ldots\right]\right)^{0.5}.$$

These advances in mathematics were essential to the development of techniques — spectroscopy, crystallography, and electronics — that have been applied to biology. The mathematics required to describe biological processes were readily available. The question relevant to this discussion is to what extent and in what ways did this sophisticated intellectual environment inform the practice of biology? Perhaps this is not the right perspective; Proclus Lycaeus (412–485) said: "Wherever there is number, there is beauty."

B3

Astronomy

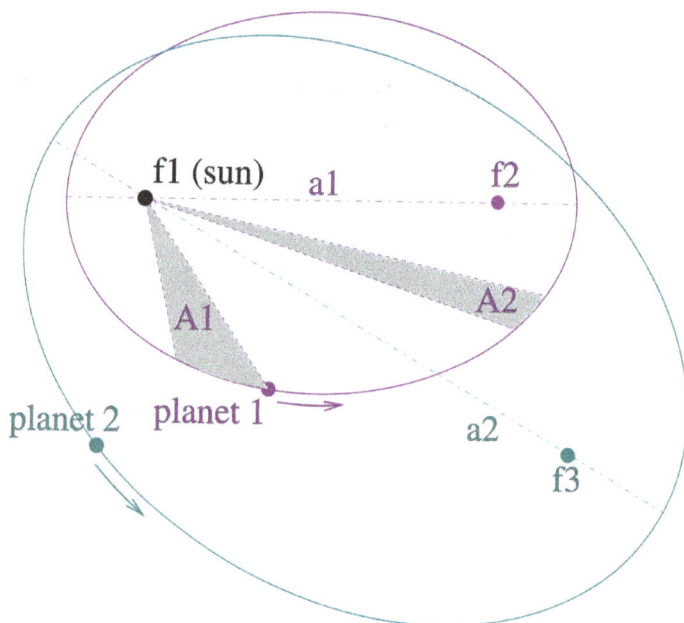

Elliptical orbits of two planets about the Sun.

There is an elegance to astronomy — no grubby worms or scat samples. Surely one is dealing with the gods and one can speak the universal language of mathematics. The Greeks, as did many earlier civilizations, made extensive observations of the heavens. Ptolemy listed 48 constellations in his *Almagest* (~150 A.D.); 30 of these had been described by the Babylonians. Today the International Astronomical Union lists 88 "asterisms."

The *Almagest* ordered the heavenly bodies — sun, moon, Mercury, Venus, Mars, Jupiter, Saturn, and the sphere of stars. Ptolemy understood that the sun and the moon were different from the planets. The stars were fixed to the inner surface of a static sphere. That was the end; there was no concept of infinity.

The publication in 1543 of the *Revolutionibus* of Copernicus is rightly cited as one of the events that laid the foundation of the scientific revolution. Copernicus proposed a heliocentric model as opposed to the generally accepted geocentric model. As will be elaborated, the Greeks had explored heliocentric models, but Aristarchus rejected them because he could not measure the predicted parallax when viewing the stars at vernal and autumnal equinoxes.

Kepler described the orbits of the six known planets as ellipses and posited the relationship: $p^2 = k \cdot a^3$, the square of the period is proportional to the cube of the major axis of the ellipse describing the orbit of that planet about the sun. Galileo was the first to employ a 3× (later 30×) lens in his telescope. He elaborated on Kepler's heliocentric model in his *Dialogue Concerning the Two Chief World Systems* (1638). He observed spots that drifted across the surface of the sun, an unanticipated imperfection. Newton in the *Principia* (1687) rationalized Kepler's laws with the assumption that the gravitational force $= k \cdot m_1 \cdot m_2 \cdot d^{-2}$; and, more important, that the same laws applied to the heavens as to earth.

Thales (~624–~546 B.C.) is regarded as the first Greek astronomer; although none of his writings have survived. About 550 B.C., he predicted a solar eclipse. Philolaus (~480–~385 B.C.) referred to the Earth revolving about a "central fire." He defined a solar year as $365\frac{1}{2}$ days.

Aristotle (384–322 B.C.) reasoned: "... ten is a pure number because there are ten bodies in the sky... Only nine — mercury, venus, earth, mars, jupiter, saturn, moon, sun, and 'the sky' — are visible so a tenth, contra-earth, is postulated." The universe is complex but finite; further, the heavens and the Earth obey different rules.

Aristarchus (310–~230 B.C.) explored a heliocentric model and tried, but failed, to measure the parallax predicted for viewing a star — vernal vs. autumnal equinox — a valid concept but an inadequate resolution. He could not imagine its great distance from the Earth, just as geologists could not imagine the great age of the Earth (Chapter B10). Man may well be the "measure(r) of all things," except distance and time.

Hipparchus (~190–~120 B.C.) is considered to be the greatest astronomer of antiquity. He developed accurate models describing the orbits of the sun and the moon, constructed chord (~trigonometric) tables, predicted solar eclipses, discovered the precession of the moon, and compiled the first comprehensive star catalog of the western world.

Claudius Ptolemy (~85–~165 A.D.) championed geocentrism in the *Almagest* (Mathēmatikē Syntaxis). He shifted the deferents of sun, moon, and each planet relative to the Earth; he introduced epicycles to explain the apparent retrograde motion of Mars. This struggle to match observation to theory nicely conforms to the crisis described by Kuhn (Chapter A11). It took only a millennium and a half for the subsequent paradigm shift.

Georg von Peuerbach (1423–1461) has been called the "father of observational and mathematical astronomy in the West." He used the "Jacob's staff" for more accurate measurement of stellar position, replaced the use of chords by sines, and calculated sine tables.

Nicolaus Copernicus (1473–1543) had studied law and medicine and worked at various day jobs, was advisor to Polish king Sigismund I the Old and physician to Duke Albrecht, as well as being a canon in the Catholic Church. He read newly available works of Cicero and Plutarch, and possibly those of Martianus Capella, which contained references to heliocentric models:

> Therefore, having obtained the opportunity from these sources, I too began to consider the mobility of the earth. And even though the idea seemed absurd, nevertheless I knew that others before me had been granted the freedom to imagine any circles whatever for the purpose of explaining the heavenly phenomena.

As early as 1520, he circulated *Commentariolus*, comments on heliocentricism, to friends. Georg Joachim Rheticus, a student of mathematics at Wittenberg, visited Copernicus in 1539. Rheticus had access to unpublished calculations of Copernicus and urged him to publish; he was involved in drafting the *Narratio prima* (First Account) in 1541. The six books of *De revolutionibus orbium coelestium* were published in 1543, the year of Copernicus' death. Andreas Osiander (1498–1552), fearing the reaction of the Church, included an unauthorized preface stating that the model was "... not physical reality, but a devise for computing." Copernicus cited the heliocentric views of Aristarchus and Philolaus. Nicole Oresme, in the 1300s in France, and Nicolaus Cusanus, in the 1400s in Germany, had discussed the motion of earth with no clerical reaction. Even so, the Catholic Church placed *De revolutionibus* on its *Index Librorum Prohibitorum* in 1616, noting that it was "... false and altogether opposed to the Holy Scriptures."

Tycho Brahe (1546–1601) had considerable family wealth and studied law at the University of Copenhagen. He was granted an estate on the island of Hven off the coast from Lund, Sweden, and built a state of the art observatory, Uraniborg. He set out to test Ptolemaic vs. Copernican systems and bought all available ephemera.

> I've studied all available charts of the planets and stars and none of them match the others. There are just as many measurements and methods as there are astronomers and all of them disagree. What's needed is a long term project with the aim of mapping the heavens conducted from a single location over a period of several years.

His subsequent observations proved invaluable to Kepler.

Giordano Bruno (1548–1600) was ordained as a Dominican priest in 1572. He left the Dominicans and Italy in 1576. He was excommunicated, he apologized; the excommunication was then revoked. He imagined his own infinite, homogeneous universe, his own mystical heliocentrism, and more important, a pantheistic, hylozoistic system, incompatible with Trinitarian beliefs. He stated his opposition to Copernicus (about 1584): "...the opinion of Copernicus that the earth did go round, and the heavens did stand still; whereas in truth it was his own head which rather did run round, and his brains did not stand still." Bruno failed to master basic interpersonal skills and was excommunicated by the Lutherans, in 1590. He applied for the chair of mathematics in Padua, in 1591, prior to Galileo. He was arrested in Venice in 1592 and transferred to Rome the next year. After a seven-year trial he was found guilty of many offenses, variations on the themes of "Holding opinions contrary to the Catholic Faith and speaking against it and its ministers" and of "Claiming the existence of a plurality of worlds and their eternity." The Inquisitor, Cardinal Bellarmine, demanded full recantation; Bruno refused. Pope Clement VIII denied his appeal. All of his works were placed on the *Index Librorum Prohibitorum* in 1603. Bruno was gagged, tied to a pole naked, and burned at the stake, in Campo de' Fiori, 17th February, 1600. Centuries later, in 2003, John Paul II expressed his "profound sorrow." The community was well aware of Bruno's fate.

Johannes Kepler (1571–1630) taught mathematics and astronomy at the University of Graz in 1594. He assisted Tycho from 1600 until his death in 1601. Kepler had access to the best observations then available. Like many of his contemporaries he was also interested in astrology and alchemy (Chapter B5). His *Mysterium cosmographicum* (1596) defended the Copernican system. He explored many models, initially five Platonic solids — tetrahedron, cube, octahedron, dodecahedron, icosahedron — inscribed and circumscribed by spherical orbs to accommodate Mercury, Venus, Earth, Mars, Jupiter, and Saturn. He refused to convert to Catholicism; he and his family were banished from Graz. Tycho invited Kepler to visit him in Prague in 1600, a year before Tycho's death. Kepler succeeded him as imperial mathematician and had full access to Tycho's observations. He first circulated his "correct" model in 1605:

1) "The orbit of every planet is an ellipse with the sun at one focus," after Copernicus.
2) "A line joining a planet and the sun sweeps out equal areas during equal intervals of time." Effectively, conservation angular momentum to be elaborated by Newton.
3) "The square of the orbital period, p, of a planet is directly proportional to the third power of the semi-major axis, a, of its orbit. Moreover, the constant of proportionality has the same value for all planets." $p^2 = k \cdot a^3$. $k = 2.97473 \cdot 10^{-19}$ $s^2 m^{-3}$ (s = sidereal year).

These ideas were elaborated in *Astronomiae Pars Optica* (1604), *Astronomia nova, Harmonice mundi* (1619), and *Epitome of Copernican astronomy: Mysterium*, 2nd ed. (1621).

Galileo Galilei (1564–1642) built several telescopes with lenses, 3× (~1605), to 30× (~1625). He described lunar mountains and craters, identified the four largest satellites of Jupiter (a "solar system" in miniature), and observed spots moving over the surface of the sun. He did not accept Kepler's elliptical orbits, but considered the circle "perfect." Cardinal Bellarmine ordered him not to "hold or defend" heliocentrism in 1616. Although the Catholic church had prohibited the advocacy of heliocentrism, Galileo printed the *Dialogue Concerning the Two Chief World Systems* in 1632. He was ordered to appear at the Holy Office in Rome, where he argued that heliocentrism was not contrary to the Scriptures. He stood trial on suspicion of heresy in 1633 and was convicted. His sentence included: He must recant his heliocentric ideas. He was ordered to be imprisoned; this was later commuted to house arrest. His *Dialogue* was banned and publication of all of his work was forbidden; they were removed from the *Index* in 1835. After receiving his sentence Galileo is quoted, perhaps apocryphally, to have muttered: "Eppur si muove"(And yet, it moves).

Johannes Hevelius (1611–1687) was Ratsherr, 1651, later mayor of Gdańsk (Danzig). One of the perks was that he could mount telescopes, initially with no lenses, on the roof of the Rathaus in 1641. He ultimately built a tube, with wire support, telescope on the ground with a 45 m focal length. He described lunar libration and detailed topography; he discovered four comets orbiting the sun, each with its own parabolic path. At the request of Henry Oldenburg, secretary of the Royal Society of London, Edmund Halley (see below) visited his fellow member of the Royal Society in Gdańsk in 1679 to confirm these observations; indeed it was the mechanism, not the lens, that provided the accuracy and contrast. Elisabeth, Hevelius' widow, published two of his works posthumously; she is regarded as the first female astronomer.

Isaac Newton (1642–1727) published his *Principia* with moral and financial support from Halley in 1687. He rationalized the elliptical orbits proposed by Kepler. He assumed an inverse square law of gravitational attraction, defined acceleration (dv/dt) in terms of his fluxions, and described, in effect, the conservation of momentum. More important, he argued the universality of these concepts. He brought the heavens to Earth, or Earth to the heavens, depending on one's perspective (Chapter B4). "Gravity explains the motions of the planets, but it cannot explain who set the planets in motion. God governs all things and knows all that is or can be done." His *Hypothesis of Light* (1675) and *Opticks* (1704) demonstrated that white light is composed of different "wave lengths," even though he interpreted light in terms of particles (Chapter B8). He devoted as much time to alchemy (Chapter B5) as to astronomy.

Edmund Halley (1656–1742) cataloged 341 southern stars in *Catalogus stellarum australium* (1679) following his visit to St. Helena. Whether Napoleon appreciated his famous predecessor remains unrecorded. Halley was appointed Astronomer Royal in 1720; he had observed his namesake comet in 1682, noted previous sightings (1456, 1531, 1607) and correctly predicted the next passage in 1758.

Frederick William (Friedrich Wilhelm) Herschel (1738–1822) emigrated from Hanover to England in 1760, as did his sister Caroline (in 1772), to work as a musician. William served as first organist in St. John the Baptist Church, Halifax, organist of Octagon Chapel, and Director of Public Concerts, Bath. He developed a deep interest in astronomy, cast and polished his own lenses, and built some of the best telescopes of the time. In 1781 he observed a new planet (a magnitude eight object) that he proposed to name Georgium III; the community preferred "Uranus." He received the Copley Medal, was elected a Fellow of the Royal Society, and appointed "The King's Astronomer" in 1782. Caroline was equally fascinated with astronomy. She discovered eight comets and three nebulae, and she updated Flamsteed's *Atlas coelestis*, a monumental task. She developed a special bond with John, William's son, and nurtured his interest and career. In 1828 she received the Royal Astronomical Society's gold medal, becoming the first female recipient.

So, what does this have to do with biology? These achievements in astronomy surely reflected the intellectual environment and helped shape it. At a step removed, a better sense of latitude and longitude, as well as of meteorology, facilitated voyages of discovery and exploration, like the one undertaken by Captain Robert Fitzroy on the HMS Beagle in 1831.

The biota, prior to human activities, has changed over thousands of years; the Earth's geology has changed more gradually, over millions of years. Galileo noted variation and movement of sunspots, having an approximate 11-year cycle. We now know that the energy reaching the earth varies some 0.1% over this cycle. Kepler got it right, to good approximation; however, there are several important second order effects.

Milankovitch in the 1910s tried to correlate the observed or inferred changes in biota with changes in climate, i.e. global warming or cooling, and these in turn with changes in the warming of the Earth by the sun (Hays *et al.*, 1976). The angle between the Earth's rotational axis and the normal to the plane of its orbit (obliquity) ranges from 22.1° to 24.5°. It is now 23.44° and decreasing; this period is about 41,000 years. Further, this tilt precesses relative to a solar coordinate system based on "fixed stars" with a period of about 26,000 sidereal years. Although this tilt, *per se*, does not affect the total solar radiation reaching the Earth, it does have two effects. The difference in solar forcing (warming) between summer and winter in both northern and southern hemispheres is more extreme the greater the tilt. The land/water (ocean) ratio is greater in the northern hemisphere and land absorbs more heat, more rapidly than does water; hence, the difference in temperature between northern and southern hemisphere is greater with greater axial tilt.

The eccentricity of the Earth's elliptical orbit varies from nearly circular (eccentricity of 0.005) to slightly elliptical (0.058) with the mean eccentricity of 0.028 with a period of about 413,000 years. The present eccentricity is 0.017. The major axis of this ellipse also precesses relative to a fixed solar coordinate system.

The periods and amplitudes of these four parameters — axial tilt and precession; ellipticity and precession — are driven primarily by the gravitational attraction of Jupiter and Saturn and vary over time. These parameters surely affect the Earth's climate; whether they fully explain its variations remains problematic. How could one understand biology without astronomy?

B4

Mechanics

Isaac Newton (1642–1727).

Isaac Newton (1642–1727) is regarded as one of the greatest of all scientists. Although his actual work on dynamics and optics did not directly address electricity and magnetism or thermodynamics, physics up until 1900 is often referred to as "Newtonian." He postulated, as Gilbert had previously suggested for the attraction of two loadstones (magnets), an inverse second power for the gravitational force. In addition to this d^{-2} postulate, he added his three "laws":

1. Every object in a state of uniform motion tends to remain in that state of motion unless an external force is applied to it.
2. The relationship between an object's mass m, its acceleration a, and the applied force F is

$$F = m \cdot a.$$

3. For every action there is an equal and opposite reaction.

Newton's first and third laws were a reformulation of the conservation of momentum. His $F = m \cdot a$ described motion in a vacuum that offered no resistance and corrected Aristotle's idea that force is required to maintain motion in "a vacuum."

Newton rationalized Kepler's three laws of planetary motion (Chapter B3) and demonstrated that celestial bodies obey the same laws as obtain on Earth. In the centuries following his publication of *Principia* (1687), ever more complex problems in mechanics and dynamics were solved within this conceptual framework. He also invented the concept of "fluxions," an awkward form of differential calculus. The notation and formulation developed about the same time by Leibnitz is now more widely used.

Newton was aware that others had wrestled with these concepts: "If I have seen further, it is by standing on the shoulders of giants." This in no way reduces the brilliance and the impact of his formulation.

Others had contemplated forces before Newton. Avicenna (980–1037) described impetus (momentum) as being weight times velocity. Johannes Kepler (1571–1630) presented his three laws as empirical; they fit the observations:

- "The orbit of every planet is an ellipse with the sun at a focus."
- "A line joining a planet and the sun sweeps out equal areas during equal intervals of time."
- "The square of the orbital period of a planet is directly proportional to the third power of the semi-major axis of its orbit. Moreover, the constant of proportionality has the same value for all planets."

Quite enough for one scientist; he did not attempt to derive them from more basic principles.

Galileo Galilei (1564–1642) was the first to systematically study acceleration. Whether he simultaneously dropped balls of different masses from the Leaning Tower of Pisa is still debated; however, he certainly did roll balls down inclined planes to get these sorts of (idealized) results:

time	0	1	2	3	4	5
distance	0	1	4	9	16	25
dx/dt		1	3	5	7	9
d^2x/dt^2			2	2	2	2

He, in effect, described acceleration without explicitly stating its formula.

He also determined that the period of a pendulum's swing is (approximately) constant with length and does not depend on its displacement.

René Descartes (1596–1650) is best known for his *Discourse on the Method* (1637) and elaboration of the deductive method (Chapter A7). He also proposed the conservation of momentum and derived an equation describing the trajectory of a projectile: $y = x \cdot \tan\alpha - g \cdot x^2/2 \cdot v^2 \cdot \cos\alpha$.

Otto von Guericke (1602–1686) built a kolbenvakuumluft-pumpe (piston vacuum air pump) in 1657. The two halves of an evacuated sphere could not be pulled apart by a team of eight horses. From the cross sectional area and the force exerted by eight horses he could calculate the atmospheric pressure. Evangelista Torricelli (1608–1647) developed a barometer and described atmospheric pressure in millimeters of mercury, much more convenient than meters of water.

John Brehaut Wallis (1616–1703) was a mathematician of broad talents. In his *Mechanica sive De Motu, Tractatus Geometricus* presented to the Royal Society in 1668, he stated: "… the initial state of the body, either rest or motion, will persist." Christiann Huygens (1629–1695) analyzed the pendulum clock and its isochronous swing in 1657. In *Horologium Oscillatorium* (1673), he derived T (period of a pendulum with weightless string) $= 2 \cdot \Lambda \cdot (l/g)^{0.5}$ (l = length of the string; g = gravitational constant). He also noted that two pendulums suspended from the same beam have coupled motions and swing in opposite directions.

These examples in no way detract from the impact of *Philosophiæ naturalis principia mathematica* (*Mathematical Principles of Natural Science*); however, many distinguished scientists were already exploring fundamental questions of mechanics and dynamics.

Newton attended Cambridge University from 1661–1665. His mathematical gifts were recognized early, and he was appointed second Lucasian Professor of Mathematics in 1669. He spent two years at home during the plague of 1663 and developed some of the ideas to be elaborated in *De Motu Corporum* (1684) and in

Principia (1687). He also published *Hypothesis of Light* in 1675, in which he posited an ether that transmitted forces between particles of light, and *Opticks* in 1704. He assumed $F = k \cdot m_1 \cdot m_2 \cdot d^{-2}$ as privately suggested by Robert Hooke (1635–1703) and previously proposed by William Gilbert (1544–1603) to describe the force of attraction between two loadstones (magnets).

Newton emphasized that terrestrial mechanics describe celestial motion. He acknowledged that he did not address the question of how those stable orbits were first established and, more important, why force was inversely proportional to distance squared. There had been sophisticated studies of mechanics and dynamics before Newton and many more would follow. Even so, *Principia* marked a turning point in physics and was the crowning glory of the scientific revolution.

Gottfried Leibniz (1646–1716) is remembered as the co-discoverer, along with Newton, of calculus and of the notation that we use today. He distinguished *vis viva* (energy) $m \cdot v^2$ from momentum, $m \cdot v$ (mass times velocity). The Bernoullis (Jakob, 1654–1705, brother Johann, 1667–1748, and Jakob's son Daniel, 1700–1782) from Basel, Switzerland all made important contributions to the application of mathematics to problems of mechanics and dynamics. Johann, as well as Leibnitz and Huygens, solved the problem of the free hanging chain, a catenary, in 1691: $y = a \cdot \cos h(x/a)[a = $ vertical displacement at the midpoint]. They also derived early formulae for hydrodynamics.

Leonhard Paul Euler (1707–1783) was also born in Basel. He held professorships in St. Petersburg, Russia, and at the Berlin Academy. He developed imaginary numbers and their application to optics with representation of amplitude and phase in the complex plane (Chapter B2). Joseph-Louis Lagrange (1736–1813) became professor of geometry at Turin in 1755. He published *Theorie des Fonctions Analytiques* in 1764 and *Mécanique Analytique* in 1788. He addressed the three-body problem for the earth, sun, and moon and discovered stationary solutions, so called Lagrangian points, in 1772.

Pierre-Simon Laplace (1749–1827) became an instructor in the École Militaire in Paris. He published *Théorie du Mouvement et de la Figure Elliptique des Planets* in 1784 and *Exposition du Système du Monde* and the *Méchanique Céleste* in five volumes (1799–1825). He proved the stability of the entire solar system.

Gaspard-Gustave de Coriolis (1792–1843) was a professor at École Polytechnique. He published *Calcul de l'Effet des Machines* (1829), *Sur le Principe des Forces Vives dans les Mouvements Relatifs des Machines* (1832), and *Sur les Équations du Mouvement Relatif des Systèmes de Corps* (1835). Jean Bernard Léon Foucault (1819–1868) suspended a large pendulum in the Panthéon, Paris, in 1851 and observed its precession over the course of a day. He noted that there should be no precession at the Equator and a 360° precession at the poles.

These few brief sketches should capture the sophistication of mechanics and dynamics at the time of Darwin. How did the intellectual achievements of physics compare with those of biology? In what ways did physics affect biology?

B5

Alchemy

Alchemist by Joseph Wright.

By necessity one tells the stories of chemistry — alchemy, phlogiston, periodic table, organic chemistry, and biochemistry — in sequence. However, both periods and paradigms overlap.

"Alchemy" is unfortunately now used as a term of derision. Its practitioners were a heterogeneous lot. There were no recognized publications, no defined admission to the profession. Some sought a magic elixir, a philosopher's stone. They spoke of earth, air, fire, and water. However, they purified and characterized S, Hg, Au, Ag, Pb, Sn, Cu, Fe, and As, seemingly without addressing the concept of an element. They understood the opposite natures of acids and bases. They borrowed many techniques from Islamic science.

The alchemists worked within a Hermetic, as opposed to a Platonic, tradition. A nominal goal was the transmutation of base metals to gold, i.e. chrysopoeia (Gr. khrusōn, gold, and poiēin, to make). In contrast, Paracelsus wrote: "Many have said of Alchemy, that it is for the making of gold and silver. For me such is not the aim, but to consider only what virtue and power may lie in medicines."

It would be difficult to describe the conceptual framework, or paradigm, in which alchemists worked. Nonetheless, they were hardly "pre-science" in the Kuhnian sense. They made many technical advances and laid the foundation for the chemistry and metallurgy that followed in the late 1600s, the scientific revolution.

It was much more difficult to establish a conceptual framework, or paradigm, for the study of chemistry than it was for astronomy or mechanics. The challenge of the necessary mathematics is well appreciated, as is the significance of Newton's *Principia*. However, one could at least see the planets and the rolling balls of Galileo (Chapter B4). Getting a handle on chemistry required the development of more complex equipment and protocols, as well as non-intuitive concepts.

Aristotle (384–322 B.C.) reasoned that different proportions of qualities — hot, cold, wet, dry — imparted different properties to substances — earth, air, water, and fire. This was a difficult concept to quantify or to subject to experiment. He endorsed the then-current idea that heating the appropriate mix of mercury and sulfur would yield a specific metal, as would alchemists many centuries later. Muhammad ibn Zakariyyā al-Rāzī (∼900) as well as Robert Boyle (1627–1691) accepted the prevailing belief in transmutation of metals. There were various statutes against multiplying gold and silver, reflecting the concern that this would debase the currency. One can consider the development and quiet demise of alchemy in this context.

Hermes Trismegistus (Mercurius ter Maximus), "thrice great Hermes," was a wise pagan god who combined characters of Corpus Hermeticum of Hermetica and the Egyptian god Thoth. The authorship(s) of the "hermetic tradition" — alchemy, magic, and astrology — remains unknown. Kepler and Newton, among many others, were intrigued by these ideas.

If men could but learn to read these divine signs, they could penetrate the mysteries of creation. And once they knew the divine language, they could hope to use it to gain control of Nature. Faust, not Galileo, was the hero in this view of the scientific revolution. Knowledge comes from the study of Nature, but only after one has been initiated into Nature's mysteries by the mastery of the "great books of hermeticism." Alchemy and astrology, alike, are branches of Truth, if one but knows the way.

Abu Musa Jābir ibn Hayyān (Geber) ($\sim$721–$\sim$815) wrote over a hundred books or pamphlets, many of which dealt with alchemy, e.g. *al-Zuhra* (*Book of Venus*) and *Kitab al-Ahjar* (*Book of Stones*). "The first essential in chemistry is that you should perform practical work and conduct experiments, for he who performs not practical work nor makes experiments will never attain the least degree of mastery." He knew that aqua regia, HCl and HNO_3, dissolves gold. He, like Aristotle, maintained that metals differ because of "different proportions of sulfur and mercury in them." He developed rust proofing of steel, dyeing and waterproofing of cloth, and tanning of leather. He characterized the "metals" — gold (Au), silver (Ag), lead (Pb), tin (Sn), copper (Cu), iron (Fe) — and noted that "... a certain quantity of acid is necessary in order to neutralize a given amount of base." Much of Geber's writing was done in code, hence the term "gibberish" (Chapter A4).

Paracelsus, nicknamed Philippus Aureolus Theophrastus Bombastus von Hohenheim (1493–1541), was born in Einsiedeln and attended the University of Basel, where he studied medicine. As an itinerant physician and mine analyst, he traveled extensively, including trips to Russia and China. Paracelsus offered an alternate view: "Many have said of Alchemy, that it is for the making of gold and silver. For me such is not the aim, but to consider only what virtue and power may lie in medicines."

He coined the term "zinke" for the sharp points on zinc crystals. For a year (1536), he was chair of medicine at Basel but managed to antagonize his colleagues. He wrote in his *Die grosse Wundartzney* (*The Great Surgery Book*): "Alle Ding' sind Gift und nichts ohn' Gift; allein die Dosis macht, dass ein Ding kein Gift ist." "All things are poison and nothing is without poison, only the dose permits something not to be poisonous."

> This is the way that nature proceeds with us in God's creatures, and follows from what I have said before, nothing is fully made, that is, nothing is made in the form of ultimate matter.

Instead all things are made as prime matter and subsequently the *vulcanus* goes over it and makes it into ultimate matter through the art of alchemy. The *archeaus,* the inner *vulcanus,* proceeds in the same way, for he knows how to circulate and prepare according to the pieces and distribution, as the art itself (alchemy) does with sublimation, distillation, reverberation, etc. For all these are in men just as they are in the outer alchemy, which is the future of them. Thus the *vulcanus* and the *achaeus* separate each other. This is alchemy, which brings to its end that which is not come to its end, which extracts lead from its ore and works it up to lead; that is the task of alchemy. Thus there are alchemists of metals, and like alchemists who treat minerals, who make antimony from antimony, sulfur from sulfur, and vitriol from vitriol, and salt into salt. Learn thus to recognize what alchemy is, that is alone is that which prepares the impure through fire and makes it pure.

Jan Baptist van Helmont (1580–1644) studied various gases; however, he had no way to store or to analyze them. Although neither wrote much of alchemy, both Tycho (1546–1601) and Newton (1642–1727) supposedly spent time doing experiments. Significant traces of mercury have been found in snippets of hair from both men.

Johann Jochim Becher (1635–1682) was appointed professor of medicine at Mainz in 1657. He published his comprehensive *Oedipum Chemicum* and *Thier- Kräuter- und Bergbuch,* in 1663, and, after visiting mines in Wales and Scotland, *Chymischer Glücks-Hafen, Oder Grosse Chymische Concordantz und Collection, Von funfzehen hundert Chymischen Processen (Chemical Harbor of Joy, or Great Chemical Concordance and Collection, Fifteen Hundred Chemical Procedures)* in 1682. He proposed a reformulation of earth, water, air, and fire. Earth and water were okay. Air and fire were subsumed under *terra mercurialis* (metallicity, fluidity, subtility, and volatility) and *terra lapidea* (fusibility). He introduced a fifth basic, *terra pinguis* (oily, sulphurous, combustible, released with burning). One can empathize with his struggle to make sense of an extraordinarily complex mass of data, with a rich overlay of mysticism.

Georg Ernst Stahl (1660–1734) was chair of medicine at Halle from 1694–1716 and personal physician to King Friedrich Wilhelm I of Prussia. He studied Becher's works and renamed *terra pinguis* "phlogiston," the subject of the next chapter.

B6

Phlogiston

Antoine Lavoisier

(1743–1794)

The concept of the four constituents — earth, air, water, and fire — provided a framework for rationalizing knowledge of what is today called chemistry, geology, and hydrology. Air and water might reasonably have been considered elements; they are abundant and seemingly indivisible. Earth is obviously a dirty mixture and, most confusingly, fire is really a process or property. Becher in 1669 wanted to extend or expand this concept, adding a fifth constituent, *terra pinguis* (oily earth). Stahl in 1731 renamed it phlogiston and elaborated on its properties. In essence it was the material postulated to be driven out of a metal when it was roasted or calcinated, or what we would now call oxidized.

Several chemists found that some substances increase in weight when roasted, i.e. obtained a layer of oxidized metal; they were forced to assign a negative weight to phlogiston in order to balance the scales. Various ad hoc attributes were added to phlogiston in order to rationalize these results. Boyle described fire particles driven out of metals, consistent with fire being a basic constituent.

This struggle faced by the chemists provided Thomas Kuhn with one of his examples of a paradigm shift. As Herbert Butterfield observed in *The Origins of Modern Science* (1957): "... where on cannot escape an anomaly, and the theory has to be tucked and folded, pushed and pinched, in order to make it conform with the observed facts."

Often an "incorrect" paradigm is better than none in that it stimulates experiments and provides some cohesion to the discipline. Whether the concept of *terra pinguis* advanced or retarded chemistry is still debated.

Without an explicit rejection of the four, or five, elements, or without involvement in the phlogiston saga, many chemists were, at the same time, characterizing more elements and comparing their fundamental properties. The ordering of these elements laid the foundations for the periodic table (Chapter B7).

The ancients had noted: "... when anything burns something of its substance streams out of it ... the original body being reduced to more elementary ingredients." Aristotle (384–322) referred to a sulfurous element of fire released during combustion. Francis Bacon (1561–1626) suggested that heat reflected the motion of tiny particles. Jan Baptista van Helmont (1579–1644) wrote of a single gas, actually another form of water, with fumes as contaminations that changed its properties; he did not appreciate that there were different gases or different elements.

As elaborated in Chapter B5, Becher proposed a reformulation of earth, water, air, and fire. Earth and water were O.K. Air and fire were subsumed under terra mercurialis (metallicity, fluidity, subtility, and volatility), terra lapidea (fusibility), and terra pinguis (oily, sulphurous, combustibility, released with burning). One can empathize with his struggle to make sense of an extraordinarily complex mass of data and mysticism.

Georg Ernst Stahl (1659–1734) was chair of medicine at Halle (1694–1716) and personal physician to King Friedrich Wilhelm I of Prussia. He was influenced by Becher's works and renamed *terra pinguis*, phlogiston.

Robert Boyle (1621–1697) noted the increase in weight upon calcination (oxidation) of metals, and ascribed this to an "... increase in weight due to insinuation of fire particles, following removal of phlogiston." Joseph Freind, professor of chemistry at Oxford in 1712 lamented:

> Chemistry has made a very laudable progress in Experiments; but we may justly complain, that little Advances have been made towards the explication of 'em ... Nobody has brought more Light into this Art than Mr. Boyle ... who nevertheless has not so much laid a new Foundation of Chemistry as he has thrown down the old.

Mikhail Vasilyevich Lomonosov (1711–1765), co-founder of Moscow State University, "repeated" Boyle's experiment and wrote in 1756: "Today I made an experiment in hermetic glass vessels in order to determine whether the mass of metals increases from the action of pure heat. The experiment demonstrated that the famous Robert Boyle was deluded, for without access of air from outside, the mass of the burnt metal remains the same."

Joseph Black (1728–1799) "discovered" nitrogen in 1772 as the residual air (N_2 and CO_2) after burning. He explained his findings in terms of phlogisticated air. Henry Cavendish (1731–1810) in *On Factitious Airs* (1766) "discovered" inflammable air, i.e. hydrogen. He measured its density and noted that on combustion it formed water. He described it as water deprived of phlogiston and oxygen as "phlogisticated water." Carl Wilhelm Scheele (1742–1786) asked: What is air? Initially he answered that all airs are fundamentally the same; they differ only in the presence of different trace factors. He also proposed the correspondence between combustion and respiration and demonstrated that several materials cannot burn in a vacuum. He discovered oxygen prior to Lavoisier but published only later.

Antoine Lavoisier (1743–1794) proposed the existence and characteristics of "anti-phlogiston," i.e. oxygen, in 1779. He is, quite rightly, credited with characterizing oxygen and sounding phlogiston's death knell, in *Réflexions sur la phlogistique* (1783). As described, prior to Lavoisier there were results that did not fit the paradigm; it was modified, e.g. negative weight, to accommodate these new observations. An alternative, oxygen, was presented; many older scientists clung to the old paradigm. Lavoisier clashed with Jean-Paul Marat (but not over phlogiston) and was guillotined. Lagrange lamented: "It took them only an instant to cut off that head, but France may not produce another like it in a century."

James Bryan Conant told this story in detail in *The Overthrow of Phlogiston Theory: The Chemical Revolution of 1775–1789* (1950). His protégé, Thomas Kuhn, chose this as one of the examples to develop the concept of a scientific revolution. Kuhn not only elaborated on the phlogiston story, he, to a great extent, structured his discussion of a paradigm shift on this example. There were several older scientists, for instance Priestly, who resisted the shift.

Joseph Priestly (1733–1804) fired hydrogen and oxygen in a closed vessel and noted that its walls were wetted. He published original observations on many types of air: "nitrous air" (nitric oxide, NO); "diminished" or "dephlogisticated nitrous air" (nitrous oxide, N_2O); "vapor of spirit of salt" or "marine acid air" (anhydrous HCl); "alkaline air" (NH_3); and "dephlogisticated air," an isolated gas in which candles burn brighter (O_2). He published the *Doctrine of Phlogiston Established and the Composition of Water Refuted* in 1800.

J. Ellicott in 1780 argued that the presence of phlogiston in a body "... weakened the repulsion between the particles and the ether ..." thereby "... diminishing their mutual gravitation." Daniel Rutherford (1749–1819), following the suggestion of Joseph Black, isolated nitrogen in 1772. He captured "noxious" or "phlogisticated" air, CO_2, and demonstrated that a candle is extinguished and a mouse dies when given only carbon dioxide.

Becher's proposal of *terra pinguis* was quite reasonable. Its evaluation and rejection certainly advanced chemistry and provided insights into physiology.

B7

Periodic Table

Periodic Table of the Elements

The development of the periodic table can be analyzed from several perspectives. The concept of many basic elements — as opposed to just earth, air, fire, and water, as well as *terra pinguis* — represented a fundamental shift of paradigm; however, no one conceded defeat or declared victory. The accumulation of observations, or experiments, leading to a generalization might be cited as an example of inductive logic, after Bacon. Perhaps most relevant to biology was the implicit or explicit search for the "correct" way to organize the mass of observations. The "natural" classification of biology, which was essential for the development of "evolution by natural selection," is fundamentally different from that of chemistry. The periodic table summarizes a vast amount of information and permits one to make important predictions. The phylogenies of biology summarize vast amounts of information and permit one to make important historical interpretations. Organizing the characteristics of organisms in a hierarchical table or the characteristics of elements in a dendrogram ain't a' goin' to work.

By 1750, sulfur, mercury, gold, silver, lead, tin, copper, iron, and arsenic had already been (partially) characterized. Lavoisier named "oxigen" and sounded the death knoll of phlogiston. One infers that his work inspired others to characterize nitrogen, hydrogen, chlorine, barium, etc. Each element has its own story and hero(s). All are fascinating; a few of the stories will be summarized below. By 1850, the list of elements had grown to about 37. Dobereiner suggested "triades, e.g. chlorine, bromine, and iodine, with similar properties. Newlands proposed "octaves," after the seven (five whole and two half) intervals of the Western musical octave of eight notes. Mendeléev in 1869 published his table as part of a textbook. Its general format, with modifications, is still used today. Its acceptance was due, in part, to his prediction of the characteristics of several elements yet to be characterized, most notably eka-aluminum, i.e. gallium.

To make and honor the distinction between fundamental entities, e.g. elements — earth, air, fire (hardly an element), and water — and their properties — cool, dry, warm, and moist — was a significant advance. During Hellenic times 12 elements — carbon (C), sulfur (S), iron (Fe), copper (Cu), arsenic (As), silver (Ag), tin (Sn), antimony (Sb), gold (Au), mercury (Hg), lead (Pb), and bismuth (Bi) — as we now know them, had been isolated and partially characterized. For two millennia people did not explicitly address this seeming contradiction of having four "elements" vs. indivisible C, S, Fe, Cu, As, Ag, Sn, Sb, Au, Hg, Pb and Bi. It was as though these materials were all subsumed in "earth." No one explicitly explored the distinction between substances and properties. No one explicitly argued that carbon, sulfur, *et al.* were themselves fundamental elements different from "earth." The shift in paradigm just happened with little fanfare.

The alchemists (Chapter B6) and miners refined techniques to purify and characterize these dozen or so elements. However, they did not explicitly set out to find new elements. With a few exceptions, to be noted, each newly discovered element had its own hero(s) and story, as nicely summarized by Sam Kean in *The Disappearing Spoon* (2010).

This discussion focuses on the events leading to the development of the periodic table. Johann Dobereiner (1780–1849) suggested triades, e.g. Cl, Br, and I, with similar properties. By 1864, when John Newlands (1837–1898) presented his table based on the "Law of Octaves" and 1869, when Dmitri Mendeléev (1834–1907) presented the table in essentially the form that we now use, about 37 elements — H, Li, Be, B, N, O, F, Na, Mg, Al, Si, P, Cl, K, Ca, Ti, V, Cr, Mn, Co, Ni, Zn, Ge, Se, Br, Rb, Sr, Y, Te, I, Cs, Ba, Ta, W, Os, Pt, Tl — had been characterized.

Democritus ($\sim$460–$\sim$370 B.C.) postulated that "... atoms (atoma, indivisible units) and the void alone exist." He suggested that tastes reflected differently shaped atoms contacting the tongue. The relationship between atoms and elements remained ambiguous.

Carl Wilhelm Scheele (1742–1786) established his own pharmacy in Stockholm in 1776 and published *Chemische Abhandlung von der Luft und dem Feuer* (Chemical Observation and Experiments on Air and Fire) in 1777, in which he isolated "fire air" (oxygen). He did not name oxygen or publish his findings before sending his results to Joseph Priestly (1733–1804), who forwarded his own and Scheele's results to Lavoisier. Scheele also purified, characterized, and named choloine gas (Cl_2, 1772, renamed "chlorine" by Humphry Davy, see below), barium (Ba, 1774), manganese (Mn, 1774), molybdenum (Mo, 1778), and tungsten (W, 1781).

Daniel Rutherford (1749–1819) isolated nitrogen, N_2, in 1772 as well as carbon dioxide, CO_2, which he referred to as phlogisticated air. Claude Louis Berthollet (1749–1822) determined the composition of ammonia, NH_3, and established a standard nomenclature for chemical compounds. William Hyde Wollaston (1766–1828) observed dark, Fraunhofer lines in the solar spectrum in 1802. He discovered palladium (Pd) and rhodium (Rh) in mine ores in 1804.

John Dalton (1766–1844) published his *New System of Chemical Philosophy* (four vols., 1808–1827) and argued that compounds are whole number ratios of their constituent elements. Further, "... it must be presumed to be a binary one, unless some cause appear to the contrary." His initial designations of water as OH and ammonia NH were subsequently corrected. In summary:

1) Elements are made of tiny particles called atoms.
2) All atoms of a given element are identical.
3) The atoms of a given element are different from those of any other element.
4) Atoms of one element can combine with atoms of other elements to form compounds.
5) A given compound always has the same relative numbers of types of atoms.

6) Atoms cannot be created, divided into smaller particles, nor destroyed in the chemical process.
7) A chemical reaction simply changes the way atoms are grouped together.

Not bad for 1808.

Lorenzo Romano Amedeo Carlo Avogadro (1776–1856) taught ecclesiastical law at the University of Turin in 1796. He published *Essai d'une manière de déterminer les masses relatives demolécules élémentaires des corps, et les proportions selon lesquelles elles entrent dans ces combinaisons* (*Essay on Determining the Relative Masses of the Elementary Molecules of Bodies and the Proportions by which They Enter These Combinations*) in 1811 and became professor of physics at Turin in 1820. He was dismissed in 1823 for his anti-royalist views ("... very glad to allow this interesting scientist take a rest from heavy teaching duties, in order to be able to give a better attention to his researches.") His "law" states that equal volumes of different gases, at the same temperature and pressure, have the same number of molecules (1833).

Joseph Louis Gay-Lussac (1778–1850) was professor of chemistry at the École Polytechnique in 1809 and professor of physics at the Sorbonne from 1808–1832. He sent a balloon to an altitude of 6.4 km and demonstrated the (near) constant composition of air with altitude in 1804. He was co-discover of boron (B), in 1808, and of iodine (I), in 1811. He established the composition of water as 2 H · 1 O; in contemporary terminology, H_2O. He added the rule, $V = k \cdot T$ (with P constant) to $P \cdot V = k'$ of *The Sceptical Chymist* (1661), by Robert Boyle.

Humphry Davy (1778–1829) purified laughing gas, nitrous oxide (N_2O), and realized its anesthetic value, but never pursued its clinical application (Chapter C1). He isolated "oxymuriatic acid," chlorine gas (Cl_2), and iodine (I_2). He realized that some acids, e.g. hydrochloric (HCl), do not contain oxygen, expanding the original definition of Lavoisier. He is best remembered for inventing the "Davy lamp," which allowed miners to burn a candle surrounded by a wire gauze screen, oxidizing methane (CH_4) without igniting an explosion. He became president of the Royal Society in 1820.

Friherre Jöns Jakob Berzelius (1779–1848) introduced the subscript notation for compounds, e.g. H_2O, still used today and published a reasonably accurate table of relative atomic weights of most of the elements then known. He discovered silicon (Si), selenium (Se), thallium (Th), and cerium (Ce), and co-discovered lithium (Li) and vanadium (V). He made the distinction that organic compounds contain carbon and coined the term "protein," denoting a "primitive substance." He became professor at the Karolinska Institute in 1807. Friedrich Wöhler (1800–1882) studied with Berzelius after receiving his M.D. from the University of Heidelberg in 1823. He is best remembered for his synthesis of urea, $CO(NH_2)_2$, from ammonia cyanate, NH_3HOCN, in 1828 (Chapters A13 and C8). He defined organic compounds as those containing carbon and found some organic material in meteorites. He (co)-discovered beryllium (Be), aluminum (Al), yttrium (Y), titanium (Ti), and silicon (Si) in the 1850s. William Thompson, Lord Kelvin, (1824–1907) proposed his "raisins in a hot cross bun" model of the atom, with electrons in the outer portion of the bun and protons in the center.

John Alexander Reina Newlands (1837–1898), who was working at the Royal College of Chemistry, published the first periodic table based on the "Law of Octaves" (eight notes or seven intervals — five whole and two half — in a row or octave, Chapter D4) in 1865. The elements were listed in sequence of increasing atomic weight. That implied a sequence of increasing atomic number, using contemporary notation. Protons and neutrons had yet to be discovered.

Dmitri Mendeleév (1834–1907) was professor of chemistry at the University of St. Petersburg (1863–1890). He became Director of the Bureau of Weights and Measures in 1893; his main task was to assure that vodka contained 40% alcohol by volume, first things first. Not unlike Kepler (Chapter B3) and Lavoisier (Chapter B6) he made sense of data already "published" in addition to his own experiments. He prepared a periodic table of elements, the direct precursor of those now in use, for his textbook, *Principles of Chemistry* (1868), and *The Dependence between the Properties of the Atomic Weights of the Elements* (1869). He predicted properties of "eka" elements yet to be discovered, among them eka-aluminum, gallium.

Paul Émile (François) Lecoq de Boisbaudran (1838–1912) published *Spectres lumineux, spectres prismatiques, et en longeurs d'ondes destines aux recherche de chimie minerale* (1874). He discovered gallium (Ga, a neat double entendre, "gal" (rooster) for le coq and the symbol for France) from zinc ore from the Pyrenees, with the properties of eka-aluminum predicted by Mendeléev. He also discovered samarium (Sm, 1880), dysprosium (Dy, 1886), and isolated gadolinium (Gd) in 1885.

William Ramsay (1852–1916) was chair of chemistry at University College London in 1887. He isolated argon (Ar), a contaminant in preparations of nitrogen, then neon (Ne), krypton (Kr), and xenon (Xe), thereby filling the eighth column, the noble gases, in the periodic table. He was awarded the Nobel Prize in Chemistry in 1904.

A few of these men, for example Ramsay and more recently Glen Seaborg, expressly searched for new elements. However, most of the early elements were discovered and characterized as parts of investigations of broader themes.

B8

Electricity, Magnetism, and Optics

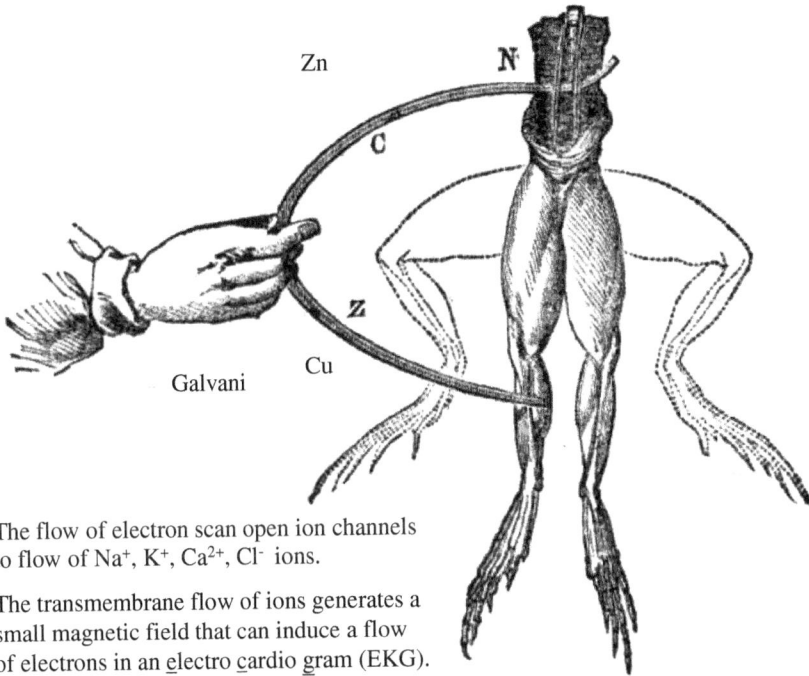

Zn   N

C

Galvani   Cu   Z

The flow of electron scan open ion channels
to flow of Na$^+$, K$^+$, Ca^{2+}, Cl$^-$ ions.

The transmembrane flow of ions generates a
small magnetic field that can induce a flow
of electrons in an electro cardio gram (EKG).

One of the most fascinating and rewarding aspects of science occurs
when several seemingly unrelated disciplines are united in a common conceptual framework.

Lodestones were described as attracting iron in 100 B.C. and
used as compasses in navigation before 1000 A.D. by the Chinese.
Benjamin Franklin, arguably the first and one of the greatest U.S.

scientists, demonstrated that lightning is in effect a discharge of electricity. Static electricity can be generated by rubbing two sorts of material, e.g. glass and fur, together, so called tribo-electricity. Lists were made of relative "donors" and "receivers" of charge. Considerable amperage and voltage, to use contemporary terms, could be stored in Leyden jars. Galvani recorded the twitching of muscles induced by an electrical potential and studied animal electricity, galvanism. The first battery was made by Volta in 1799; it was quickly improved and widely used. This bit of gadgetry provides a good example of why engineering in its broadest sense should be included in the definition and discussion of science.

Many had observed the invariant sequence of colors (red, orange, yellow, green, blue, indigo, and violet) in rainbows. Roger Bacon noted that the spectrum of colors refracted from a glass edge resembled a rainbow. Descartes stated Snell's law defining the index of refraction of media that transmitted light. Newton first passed white light through a prism and observed the rainbow of colors. He then reconstructed white light using a complementary prism. He also showed that the colors from the first prism could not be further separated. Although Newton described light in terms of particles; most accepted the description of light and its diffraction in terms of waves, as formulated by Young.

Ampère and Örsted observed that an electrical current deflects a compass and conversely that the movement of a wire through a magnetic field induces a current in the wire.

The synthesis of optics, electricity, and magnetism, as summarized in Maxwell's four equations, might be regarded as the crowning glory of 19th century Newtonian physics. Then along came folks like Roentgen, Einstein, and Bohr (as well as Monet and Renoir) and screwed up everything!

The initial investigations of magnetism, of electricity, and of optics were seemingly unrelated. This chapter summarizes the early experiments and theories in the three fields, then the insights into their interrelationships, and finally their relevance to biology.

William Gilbert (1544–1603) was personal physician to Elizabeth I and James I (VI of Scotland). His *De magnete, magneticisque corporibus, et de magno magnete tellure* (1600) summarized his investigations of magnetic stones. He suggested a d^{-2} force of attraction between magnets. This relationship was also assumed by Newton for gravitational attraction.

Otto von Guericke (1602–1686) devised an electrostatic generator and explored tribo (rub) electricity, i.e. the rank order, positive to negative of whether a material — cloth, fur, paper — tends to accumulate or to lose a few electrons when rubbed with another material. Robert Boyle (1627–1691), author of *The Sceptical Chymist* (1661), observed that electrical attraction could be propagated across a vacuum, i.e. corpuscles were not required. Benjamin Franklin (1706–1790) enjoyed a rich and varied life in politics and was the first scientist of note from the United States. He postulated distinct positive and negative charges and an "electrical fluid." He rationalized lightning rods with sharp points and famously demonstrated the electrical nature of lightning. But that all of our diplomats were so gifted. Joseph Priestley (1733–1804) is best known for his enlightenment rationalism, e.g. *Essay on the First Principles of Government* (1768). He was a member of the Lunar Society of Birmingham (1765–1813), a group of natural philosophers who met on nights of the full moon so they had better light for their late rides home. Most of his scientific achievements concerned characterizing various gases within the context of the phlogiston theory.

Luigi Galvani (1737–1798) observed that he could induce the leg muscles of dead frogs to twitch when touched with probes made of two different metals, e.g. iron and copper. Conversely, he suggested that muscle and nerve cells produce electricity. Alessandro Giuseppe Antonio Anastasio Volta (1745–1827) was professor of experimental physics at the University of Pavia in 1779, and professor of philosophy at Padua in 1815. He studied "animal electricity" as first described by Galvani and invented the voltaic pile, or battery, in 1799. It consisted of alternating sheets of metal, e.g. zinc and copper, separated by paper soaked in salt. Having a defined, reproducible source of direct electrical current opened the field to quantitative study; this is a lovely example of an advance in engineering enabling advances in pure science.

"Electricity" usually refers to the longitudinal flow of electrons, at nearly the speed of light. In contrast "impulses" in nerves result from the flow of ions — Na^+, K^+, Ca^{2+}, and Cl^- — across a membrane, thereby depolarizing the adjacent region of the membrane and initiating further trans-membrane flow of electrons. The resultant longitudinal nerve impulse does not involve longitudinal passage of either electrons or ions. It travels at 5–$120\,m \cdot s^{-1}$. An electric current can initiate this depolarization resulting in a nerve impulse. The trans-membrane flow of ions generates an electric field that can be detected in, for instance, an electrocardiogram (EKG).

René Descartes (1596–1650) defined the refractive index, $n = \sin i / \sin r$ (i is the angle of incidence; r is the angle of the refracted ray) and deduced the angular radius of a rainbow, 42°. Christiaan Huygens (1629–1695) read his *Treatise on Light* to the Royal Society in 1679: "It is inconceivable to doubt that light consists in the motion of some kind of matter." "This is assuredly the mark of motion, at least in the true Philosophy, in which one conceives the causes of all natural effects in terms of mechanical motions." Huygens actually described longitudinal waves and did not consider wavelength and color. Isaac Newton (1643–1727) published *Hypothesis of Light* (1675), and *Opticks* (1704). He described how white light could be decomposed into its constituent colors by passage through a prism. An inverted prism could combine these colors into white light; passage of one "color" though a second prism produced no further decomposition. He posited ether to transmit forces between the particles of light. The seeming conflict between wave and particle descriptions of light may reveal more of human thought processes than of fundamental physics.

Thomas Young (1773–1829) was a true polymath. In *The Last Man Who Knew Everything* (2007), Andrew Robinson wrote:

> Young probably had a wider range of creative learning than any other Englishman in history. He made discoveries in nearly every field he studied. In addition to the initial decipherment of the Rosetta Stone, he studied vision. In his first paper to the Royal Society at age twenty he wrote, "It is well known that the eye, when not acted upon by any exertion of the mind, conveys a distinct impression of those objects only which are situated at a certain distance from itself; that this distance is different in different persons, and that the eye can, by volition of the mind, be accommodated to view other objects at a much lesser distance; but how this accommodation is effected, has long been a matter of dispute, and has not been satisfactorily explained."

Young presented three hypotheses: the curvature of the cornea changes; the length of the eyeball changes; the shape of the crystalline lens changes. He then demonstrated that the lens changes. He argued that the eye detects three primary colors. Maxwell wrote: "Thomas Young was the first who, starting from the well-known fact that there are three primary colors, sought for the explanation of this fact, not in the nature of light but in the constitution of man."

Hans Christian Örsted (1777–1851), professor of physics at the University of Copenhagen in 1820, observed that an electric current deflects a magnet. He produced aluminum metal by electrophoretic deposition from a solution of aluminum salt in 1825. André-Marie Ampère (1775–1836) held several professorships in physics, chemistry, and mathematics. He heard of Örsted's discovery in September, 1820, and a week later read a paper and made a demonstration to the French Academy that parallel wires (*l* long and *r* apart) that carry parallel currents attract and anti-parallel wires repel with force $F = k \cdot I_1 \cdot I_2 \cdot r^{-1}$. One ampere of current is one coulomb / second past a fixed point and exerts a magnetic force of 2×10^{-7} newtons per meter.

Ampère was an encyclopedist with a modest overview, *Essai sur la philosophie des sciences, ou exposition analytique d'une classification naturelle de toutes les connaissances humaines* (*Essay on the Philosophy of the Sciences, or an Analytical Exposition of a Natural Classification of all Human Knowledge*). In 1827 he published his magnum opus, *Mémoire sur la théorie mathématique des phénomènes électrodynamiques uniquement déduite de l'experience* (*Memoir on the Mathematical Theory of Electrodynamic Phenomena, Uniquely Deduced from Experience*).

Michael Faraday (1791–1867) had little formal education and was apprenticed to a local bookbinder. He attended lectures by Humphry Davy in 1812, became his secretary, then lab assistant, and even valet on a tour through Europe. He excelled as an experimentalist and twice refused the Presidency of the Royal Society. He delivered 19 annual Christmas lectures, including *The Chemical History of a Candle*, at the Royal Institute in London. He determined the charge of a mole of electrons (96,485 coulombs) and first demonstrated the interaction of light with a magnetic field, stating: "I have at last succeeded in illuminating a magnetic curve or line of force and in magnetising a ray of light."

Hermann Ludwig Ferdinand Helmholtz (1821–1894) had an enormous range of achievements. He argued the conservation of energy in *Über die Erhaltung der Kraft* in 1847 and cited the studies of Mayer and Joule. He presented a physiological theory of music, *Die Lehre von den Tonempfindungen* (*On the Sensations of Tone as a Physiological Basis for the Theory of Music*, 1863). He predicted and demonstrated electromagnetic radiation.

Armand Hippolyte Louis Fizeau (1819–1896) and Jean Bernard Léon Foucault (1819–1868) measured the speed of light with and against a stream of flowing water and established that the index of refraction could be deduced from the slower speed of light in any medium relative to its speed in a vacuum (1849). Their "correction" to the speed of light was one of the observations that inspired Einstein's theory of relativity. They argued that the "... length of a light wave be used as a length standard ..." in 1864.

Gustav Robert Kirchhoff (1824–1887) was chair of theoretical physics at the University of Berlin. His three laws of spectroscopy are: a hot solid emits a continuous spectrum; a hot gas emits discrete spectral lines; and a hot solid, surrounded by cool gas emits a continuous spectrum with gaps (absorption lines of the gas). He and Bunsen, of burner fame, discovered cesium and rubidium in the sun's spectrum in 1861 (Chapter B7).

James Clerk Maxwell (1831–1879) was appointed professor of physics at Cambridge in 1871, where he developed the Cavendish Laboratory. In his landmark *A Dynamical Theory of the Electromagnetic Field* (1864), he summarized: "The agreement of the results seems to show that light and magnetism are affections of the same substance, and that light is an electromagnetic disturbance propagated through the field according to electromagnetic laws." Maxwell described light as electromagnetic waves of oscillating electric and magnetic fields at $310{,}740{,}000\,\text{m} \cdot \text{s}^{-1}$ (today's best value, $299{,}792{,}458$). His four equations are still the starting point for calculations in "*E & M*":

1.

$$\int E \cdot dA = q \cdot e_0^{-1} \text{ (total flux of electric field out of}$$

$$\text{a closed surface)}$$

$$= \Sigma \text{ enclosed charge, } q, \cdot e_0^{-1}$$

$$dA = \text{vector} \perp \text{to patches covering the surface.}$$

2.

$$\int B \cdot dA = 0 \text{ (closed surface)}$$
$$B \text{ magnetic field}$$

no magnetic monopoles, net magnetic flux $= 0$

$$\blacktriangledown \cdot E = 0.$$

3.

$$\int E \cdot ds = -d \, (\int B \cdot dA)/dt$$
$$\blacktriangledown \cdot E = Q \cdot e_0^{-1}(\text{Gauss' Law})$$
$$d/dt \text{ current flow over time}$$
$$ds = \text{vector} \parallel \text{to curve}$$
$$\blacktriangledown \cdot B = 0.$$

4.

$$\int B \cdot dS = m_0(\int J \cdot dA + e_0 d/dt(\int E \cdot dA))(\sim\text{Amp's Law})$$
$$\text{(path integral over surface spanning paths)}$$
$$\int J \cdot dA = -dq/dt \, J = \text{current density}$$
$$d/dt(\int E \cdot dA) = \text{displacement current}$$
$$\blacktriangledown \times B = m0J$$

electric and magnetic fields: $v = 3.10^8 \, \text{m} \cdot \text{s}^{-1}$.

Einstein was unreserved, calling Maxwell's work "… the most profound and the most fruitful that physics has experienced since the time of Newton."

Much of biology from ethology to physiology is informed by "E & M."

B9

Thermodynamics

$$S = k \cdot \ln W$$
$$W = N! \, / \, \textstyle\prod_i N_i \,!$$
$$k = 1.3806505 \cdot 10^{-23} \, \text{J} \cdot \text{K}^{-1}$$

Tombstone of
Lvdwig Boltzmann
(engraved).

The early development of thermodynamics involved at least five new ideas. The fire — of earth, air, fire, and water — confounded the substance that was hot with the property of warmth. The distinction of substance from property occurred without anyone explicitly describing this shift. Gradually folks developed the idea that heat could be measured and that different sources of heat could be compared. The third step was the concept of energy and the realization that heat was only one of its many manifestations. Other forms of energy — chemical, electrical, and especially kinetic ($E = m \cdot v^2$) — were measured and equated. This appreciation of equivalence led to the concept of the conservation of energy.

The fifth idea was the development of the concept of entropy by Carnot, Rankine, Clausius, and many others. Boltzman presented one of the more creative ideas in science, that entropy is in effect a measure of the number of different states available to the system and that this in turn is a measure of the disorder of the system. As engraved on his tombstone, $S = k \cdot \ln W$.

As though this were not enough, Claude Shannon in 1947 published one of the most important, and initially overlooked, papers in all of science equating the entropy of a message with its information content and in turn distinguishing information from meaning. All this because some engineers wanted to build more efficient steam engines and others wanted to optimize their telegraph and radio transmissions.

Contemporary biologists may ask whether the information content of the genome, as defined by Shannon, is in any meaningful way related to functional genomics, proteomics, etc.

The development of thermodynamics can be considered in terms of five concepts:

1) The property of warmth can be distinguished from the characterization of substance — earth, air, and water; later from the characterization of elements and their compounds (Chapter B7).
2) Heat content and temperature can be measured.
3) The conservation of mass and the conservation of energy were assumed, implicitly or explicitly, by several people.
4) Energy has, in addition to heat, many manifestations — work, chemical, electrical, potential, and especially kinetic ($E = m \cdot v^2$) — that can be measured and equated.
5) Entropy can be defined as $S = k \cdot \ln W$; the entropy of an isolated system tends to a maximum over time.

We now know that if system A is in equilibrium with B and B in equilibrium with C, then A is in equilibrium with C. The "three laws of thermodynamics" are now formulated as:

1) $E_2 - E_1 = Q - W$ (conservation of energy)

E = internal energy of two states in equilibrium;

Q = heat transferred between systems;

W = work done by the system.

2) $dS/dt > 0$ the entropy of an isolated system tends to a maximum over time.

3) As a system approaches absolute zero, all processes cease and the entropy of the system approaches a minimum value. Walter Nerst (1864–1941) stated that for a perfect crystal $S \to 0$ as temperature $\to 0$ K(elvin), i.e. there is only a single atomic state.

One can follow the evolution of these ideas from the work of individual scientists. Jan Baptist van Helmont (1577–1644) and Otto von Guericke (1602–1686) both argued conservation of mass in various physical and chemical processes. Hermann Ludwig Ferdinand Helmholtz (1821–1894) proposed the conservation of energy in *Über die Erhaltung der Kraft* in 1847 and credited the studies of Mayer and Joule.

Equivalent forms of energy were explored by various men. Gottfried Leibniz (1646–1716) clarified the distinction between *vis viva* (energy) $m \cdot v^2 =$ and momentum $m \cdot v$ ($m =$ mass; $v =$ velocity), which figured in Newton's first and third laws. (Every object in a state of uniform motion tends to remain in that state of motion unless an external force is applied to it. For every action there is an equal and opposite reaction.)

Benjamin Thompson, Count Rumford (1753–1814), described the heat generated in terms of the work done and conservation of energy in boring a cannon in *An Experimental Enquiry Concerning the Source of the Heat which is Excited by Friction* (1798). John Dalton (1766–1844) argued: "I see no sufficient reason why we may not conclude that all elastic fluids (ones that can expand in all directions) under the same pressure expand equally by heat... It seems, therefore, that general laws respecting the absolute quantity and the nature of heat are more likely to be derived from elastic fluids than from other substances."

Other scientists explored ideas that ultimately were subsumed under the broad envelope of "thermodynamics." Robert Boyle (1627–1691) in *The Sceptical Chymist* (1661) proposed that the product of pressure times volume is constant — "Boyle's law." Daniel Bernoulli (1700–1782) attempted to formulate a kinetic theory of gases to explain Boyle's law, in a sense incorporating Newtonian mechanics into thermodynamics. Joseph Black (1728–1799) distinguished between latent heat, specific heat and heat capacity. He demonstrated that melting ice absorbs heat at constant temperature.

Jean Baptiste Joseph Fourier (1768–1830) derived equations for the propagation of heat in his *Théorie analytique de la chaleur* (1822). He also noted that planets lose energy, "chaleur obscure," via infrared radiation and suggested in 1824 that gases in the atmosphere might increase the surface temperature of the Earth — only 200 years ahead of this time (Chapter D5)!

Julius Robert von Mayer (1814–1878), as ship physician enroute to Jakarta, asked whether storm whipped waves were warmer than calm sea and suggested that oxidation was the primary source of usable energy for any creature, in *On the Quantitative and Qualitative Determination of Forces* (1841). "Energy can be neither created nor destroyed." He estimated that the sun should cool in 5,000 years and speculated that the impact of meteorites might keep it hot. In *Die Mechanik der Warme* (1867), he defined $R = C_p - C_v$ (heat capacity at constant pressure minus heat capacity at constant volume).

James Prescott Joule (1818–1889) with Benjamin Thompson proposed the absolute scale for temperature and wrote in 1843:

> …the mechanical power exerted in turning a magneto-electric machine is converted into the heat evolved by the passage of the currents of induction through its coils; and, on the other hand, that the motive power of the electro-magnetic engine is obtained at the expense of the heat due to the chemical reactions of the battery by which it is worked.

In retrospect so much is obvious. Yet the concept of "conservation," in the sense of an isolated system containing a fixed amount, is absolutely non-trivial and can easily be misapplied. Is the wealth of a society conserved? Or the amount of love given by parents to their children? If there are winners, must there be losers?

Gustav Robert Kirchhoff (1824–1887) described black-body radiation and proposed the law of thermal radiation in 1859 (the same year as the publication of Darwin's *Origin of Species*). His three laws of spectroscopy are: a hot solid emits a continuous spectrum; a hot gas emits discrete spectral lines; and the spectrum of a hot solid, surrounded by a cool gas emits a continuous spectrum with discrete absorption lines due to the gas. He was appointed chair of theoretical physics at the University of Berlin in 1875.

James Clerk Maxwell (1831–1879) independent of Boltzmann proposed a kinetic theory of gas, now referred to as the Maxwell–Boltzmann kinetic theory. Josiah Willard Gibbs (1839–1903) is remembered for his *On the equilibrium of heterogeneous substances* (1876), in which he described his phase rule $F = C - P + 2$ (F = degrees of freedom; C = components; P = phases).

Henri-Louis Le Chatelier (1850–1936) became professor of chemistry at École des Mines in 1887. He proposed that an isolated, chemical system will tend toward its energy minimum.

Nicolas Leonard Sadi Carnot (1796–1832) defined the efficiency of an "ideal heat engine" in *Réflexions sur la puissance motrice du feu* (*Reflections on the Motive Power of Fire*, 1824) as ($1 - T_{cold}/T_{hot}$), the temperatures of two heat reservoirs. His "caloric" later became the "entropy" of Clapeyron, Clausius, and Kelvin (1834). William John Macquorn Rankine (1820–1872) was professor of civil engineering at Glasgow University (1855–1872); he was interested in all things locomotive and proposed relationships between temperature, C_p, and C_v. He proposed a "function" in 1849 identical to the entropy of Clausius.

Rudolf Julius Emanuel Clausius (1822–1888) held professorships at the Eidgenössische Technische Hochschule in Zürich (1855), University of Würzburg (1867), and University of Bonn (1869). He presented his mechanical theory of heat, formulated the basic idea of the second law, $dS/dt > 0$, and introduced the term "entropy" in *Über die bewegende Kraft der Wärme* (*On the Moving Force of Heat*, 1850). He reformulated the Carnot and introduced his gas-kinetic model with translational, rotational, and vibrational motions as well as the mean free path in 1857. "The energy of the universe is constant." "The entropy of the universe tends to a maximum" — *Abhandlungen über die mechanische Wärmetheorie, Zweite Abteilung* (1867).

Astronomers today struggle with application of this maxim to the "Big Bang."

William Thomson (Lord Kelvin) (1824–1907) was appointed chair of natural philosophy at the University of Glasgow in 1846. He was skeptical of inter-convertibility, among other things: "... the conversion of heat into mechanical effect is probably impossible, certainly undiscovered." "... the whole theory of the motive power of heat is founded on...two...propositions, due respectively to Joule, and to Carnot and Clausius." "It is impossible, by means of inanimate material agency, to derive mechanical effect from any portion of matter by cooling it below the temperature of the coldest of the surrounding objects." Heat is "...lost to man irrecoverably; but not lost in the material world."

Ludwig Eduard Boltzmann (1844–1906) received his Ph.D. from the University of Vienna in 1866, then held a series of academic positions before returning to the university in 1902. He, as did Maxwell, developed statistical thermodynamics; the Maxwell–Boltzmann equation describes the distribution of velocities of molecules in a gas. He defined entropy, $S = k \cdot \ln W$ (as engraved on his tombstone). W (Wahrscheinlichkeit, probability) $= N!/\Pi_i N_i!$ N = microstates available to the system. $k = 1.38 \cdot 10^{-23}$ JK^{-1} (joules per degree). He could hardly have anticipated that Claude Shannon, at Bell Labs in 1947, would equate entropy with information (Chapter C16).

Biology obeys the laws of thermodynamics at both microscopic and macroscopic levels. This realization helped lay to rest vitalism and is fundamental to addressing arguments of creationism and intelligent design.

B10

Geology

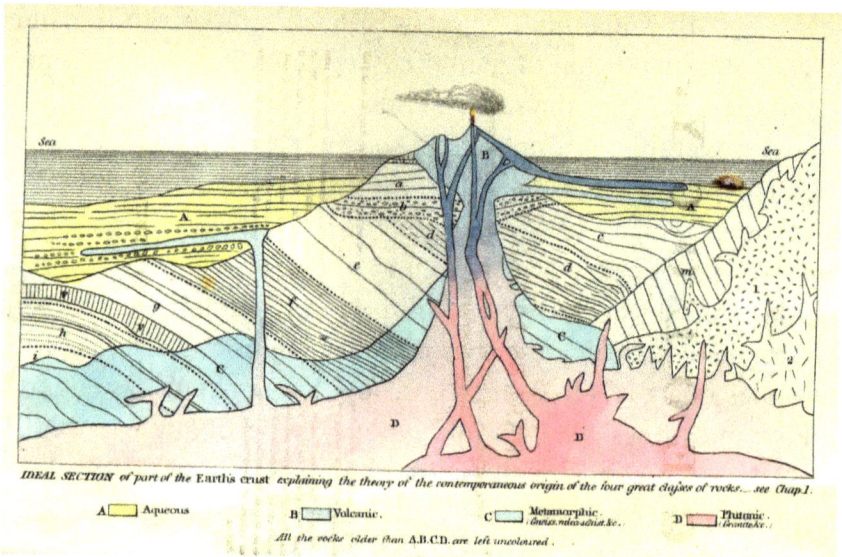

IDEAL SECTION of part of the Earth's crust explaining the theory of the contemporaneous origin of the four great classes of rocks....see Chap.1.

A☐ Aqueous.   B☐ Volcanic.   C☐ Metamorphic.   D☐ Plutonic.

All the rocks older than A.B.C.D. are left uncoloured.

Principles of Geology (frontispiece).

One might well consider the study of what lies beneath the heavens to be as significant as what lies beyond. Yet the planets and stars had a certain elegance and simplicity while the Earth was dirty and complex. The Greeks' sense of geography was recorded in their maps of the known, Mediterranean world, most famously in Ptolemy's *Geography* (~150 A.D.). Tides were measured and fossils marveled at, yet there was no unifying story or concept, except one of creation.

Georgius Agricola summarized the contemporary knowledge of metals, minerals, and mining in *De re metallica* (1556). Hutton was one of the first to record systematic observations of strata and the sorts of rocks and fossils that they contained. He recorded their tilts as well as their constancies over extended areas of Scotland and England and wrote *Theory of the Earth* (1785). Lyell wrote the widely read *Principles of Geology* (three vols., 1830—1833) and maintained that the Earth was old, perhaps very old. He mentored and championed Darwin and certainly influenced his thinking.

Two alternate theories were explored: neptunism, that rocks crystallized from salts in the oceans; and plutonism, that rocks came from volcanic eruptions. In contrast to the ideas of Copernicus and Galileo, these theories concerning the origin(s) of the Earth were not regarded as threatening by the Catholic Church.

Wegener presented his theory of continental drift in 1912. However, it was accepted in the 1960s only after the underlying mechanism of plate tectonics was elaborated. This was certainly a paradigm shift and was very quickly accepted by older members of the scientific community.

One can today cite eon, era, and period. Yet at the immediate level of human experience, when viewing the Grand Canyon or rounded granite boulders at the sea shore, it is still difficult for most of us to really appreciate the antiquity of life on Earth, let alone of the universe itself. Engaging that sense of deep time was essential for Darwin and his colleagues to grasp the infinitesimally tiny steps of evolution.

As is so often the case, practical application, in this case mining, contributed to geology just as the developing concepts of geology brought reason to mining. These studies of metals and of coal overlapped alchemy (Chapter B5); to what extent the two influenced one another is not clear.

As is usually the case, many people contributed to the development of the main concepts of geology; summaries of a few capture the general sequence and pattern. Several Greeks, among them Thales (~624–~547 B.C.), Anaximander (~610–~546 B.C.), and Herodotus (~484–425 B.C.), wrote of marine fossils found inland and suggested that land lay under water in earlier times. Georgius Agricola (Georg Bauer, "farmer" in German, 1490–1555) was the town physician in Chemnitz, a mining town in Germany. He wrote many articles on various aspects of geology. His *De re metallica* (1556) was translated into English by Lou Henry and Herbert Hoover in 1912.

Nicolas Steno (1638–1686), for whom the *ductus stenonianus* from the parotid gland is named, was a distinguished anatomist (Chapter C2). He realized that "tongue stones" are fossilized shark teeth. In his *Dissertationis prodromus* (1669), he wrote "... at the time when any given stratum was being formed, all the matter resting upon it was fluid, and, therefore, at the time when the lower stratum was being formed, none of the upper strata existed." "Strata either perpendicular to the horizon or inclined to the horizon were at one time parallel to the horizon." "Material forming any stratum were continuous over the surface of the Earth unless some other solid bodies stood in the way." "If a body or discontinuity cuts across a stratum, it must have formed after that stratum." Good introduction to a contemporary text.

Edmund Halley (1656–1742), in addition to being a distinguished astronomer (Chapter B3), made a *General Chart of the Variation of the Compass* with isogonic lines in 1701. He proposed a model of the Earth consisting of concentric shells. Georges-Louis Leclerc, Comte de Buffon (1707–1788), became a member of the French Academy of Sciences in 1734 and of keeper of the Jardin du Roi (now Jardin des Plantes) in 1739. Although not a field biologist, he wrote extensively for the general reader — *Histoire naturelle, générale et particulière* (36 vols., from 1749 on). He described species as "improved" or "degenerated" since creation and considered those, including *Homo*, of the Americas to be less vital than their European counterparts. This motivated Thomas Jefferson to write his only book, *Notes on the State of Virginia* (1785), to refute Buffon's assertions. Buffon also proposed that planets arose from comets and calculated that the Earth was about 75,000 years old based on the cooling rate of molten iron.

James Hutton (1726–1797) was a member of the Scottish Enlightenment, which included Playfair, Black, Hume, and Smith. He argued that granite penetrated the schist; hence, granite was younger and once molten — an example of the penetration of volcanic rock through pre-existing sedimentary rock. He opposed the "Neptunist" theory that all rocks precipitated from a single enormous flood. "The result, therefore, of our present enquiry is, that we find no vestige of a beginning, no prospect of an end." As a "Plutonist" he proposed that the interior of the Earth is hot. He distinguished heritable variation as exploited in animal breeding from non-heritable variation reflecting the environment.

> ... if an organised body is not in the situation and circumstances best adapted to its sustenance and propagation, then, in conceiving an indefinite variety among the individuals of that species, we must be assured, that, on the one hand, those which depart most from the best adapted constitution, will be the most liable to perish, while, on the other hand, those organised bodies, which most approach to the best constitution for the present circumstances, will be best adapted to continue, in preserving themselves and multiplying the individuals of their race.
>
> — *An Investigation of the Principles of Knowledge and of the Progress of Reason from Sense to Science and Philosophy* (vol. 2, 1794).

Friedrich Wilhelm Heinrich Alexander Freiherr von Humboldt (1769–1859) had far-ranging interests in natural history; his writings had broad appeal and influence. He noted that volcanoes occur in linear groups and that the magnetic field decreases from the poles to the Equator. He proposed that South America and Africa were once joined. His five-volume *Kosmos* (1845) attempted to unify science. Jefferson wrote: "I consider him the most important scientist whom I have met." Darwin stated: "He was the greatest travelling scientist who ever lived." "I have always admired him; now I worship him."

Mary Anning (1799–1847) lived on the coast at Lyme Regis, west of Dorchester. She, like her father, was a fossil hunter. She found the first complete skeleton of an ichthyosaur in 1811, of *Plesiosaurus dolichodeirus* in 1821, and *Pterodactylus macronyx* in 1828. She argued that these were evidence for extinction as opposed to undiscovered living species. Around 1840, she was made an honorary member Geological Society of London but, as a woman, was ineligible for regular membership.

Charles Lyell (1797–1875) was strongly influenced by Hutton. He in turn served as a mentor to Charles Darwin, who took Volume 1 of his *Principles of Geology* with him on HMS Beagle and received Volume 2 while in Brazil. Lyell supported uniformitarianism: "... the present is the key to the past." "The imagination was first fatigued and overpowered by endeavouring to conceive the immensity of time required for the annihilation of whole continents by so insensible a process." As will be discussed in Chapter C14, Lyell supported evolution, but questioned the mechanism, natural selection.

Louis Rodolphe Agassiz (1807–1873) wrote in *Etudes sur les glaciers* (1840):

> ... that great sheets of ice, resembling those now existing in Greenland, once covered all the countries in which unstratified gravel (boulder drift) is found; that this gravel was in general produced by the trituration of the sheets of ice upon the subjacent surface, etc.

William Thomson (1824–1907), in *Of Geological Dynamics* (1869), calculated that the Earth was older, 20 to 400 millon years, than others had proposed. As will be later emphasized (Chapter C14), this increasing sense of age of the Earth was essential to the development and acceptance of Darwin's theory.

Most of these chapters on physics and biology explore the intellectual environment in which Darwin worked and subsequently in which evolution by natural selection was debated. This discussion of geology is extended to continental drift and plate tectonics because it explains the correspondence of fossils between continents.

Alfred Lothar Wegener (1880–1930) noted identical fossils in similar strata separated by the Atlantic Ocean in 1911. Helmholtz had speculated that Brazil and Africa were once joined. Wegener in 1915 first explicitly proposed continental drift (*"Verschiebung der Kontinente"*) and that a supercontinent, "Pangea," had split apart about 180 million years before the present. The American Association Petrology and Geology symposium adopted a resolution refuting continental drift in 1924; because there was no mechanism (*Plate Tectonics, An Insider's History of the Modern Theory of the Earth*, ed. Naomi Oreskes, 2003). Harry Hammond Hess (1906–1969) was Professor of Geology at Princeton University (1934–1969). As captain of the USS Cape Johnson, a transport ship with sonar, he identified flat-topped submarine volcanoes, guyots, (named after Arnold H. Guyot). He subsequently had access to data from the south Atlantic that indicated symmetric changes in the direction of north–south magnetization about the south Atlantic ridge; this in effect dated the magma extruded and flowing eastward and westward. He proposed a mechanism of plate tectonics, i.e. this north–south line of extrusion pushing the American plates apart from the African and European plates, in his report to the Office of Naval Research in 1960. Within a few years the entire community accepted continental drift and the underlying mechanism of plate tectonics.

Luis W. Alvarez (1911–1988) received the Nobel Prize in 1968 for "... the discovery of a large number of resonance states, made possible through his development of the technique of using hydrogen bubble chamber and data analysis." With his son, Walter, he identified high levels of iridium, rare on Earth but in relatively higher content in asteroids, at the K(cretaceous)–T(tertiary) boundary and proposed that an asteroid impact and subsequent cooling of the Earth by dust blocking sunlight led to the demise of dinosaurs and many other fauna and flora. The site of this impact was subsequently identified at Chicxulub off the east coast of Mexico in 1990. What better example of the interaction of various disciplines, astronomy and geology, and their impact on biology!

Section C

Biology: Overview

Biology can be considered from two perspectives, reductionist and historical. The former asks about the components of a system, then how they interact, at ever higher resolution and greater detail. This is how most of physics works; biology has benefitted enormously from both the concepts and tools of physics, as outlined in the previous section. This reductionist approach to biology is closely associated with medicine and agriculture just as engineering is related to physics. The challenge to the historian is to tease out these technical and intellectual interactions, perhaps not even recognized by the practitioner.

The historical, or synthetic, approach describes how a system developed over time. Within physics only geology and astronomy incorporate this temporal dimension. The basic laws of physics are assumed to be time invariant, at least after the Big Bang. Understanding the time scale of geology transformed the way that biologists thought about evolution and ecology. Natural history emphasizes the identification and description of organisms, initially plants and animals. These ever larger compendia forced questions of variation and classification. How best can we make sense of them, beyond marveling at their beauty?

Most exciting, the reductionist and the synthetic approaches to biology complement each other evermore. Natural history incorporates DNA sequencing; molecular and developmental biology ask questions about the history of the components involved. Evolution by natural selection provides the conceptual framework for understanding life.

C1. Medicine
C2. Anatomy
C3. Physiology
C4. Cell Biology
C5. Embryology
C6. Microbiology
C7. Pharmacology
C8. Biochemistry
C9. Neurobiology
C10. Botany
C11. Genetics
C12. Paleontology
C13. Systematics
C14. Evolution
C15. Race
C16. Information
C17. Origin of Life

C1

Medicine

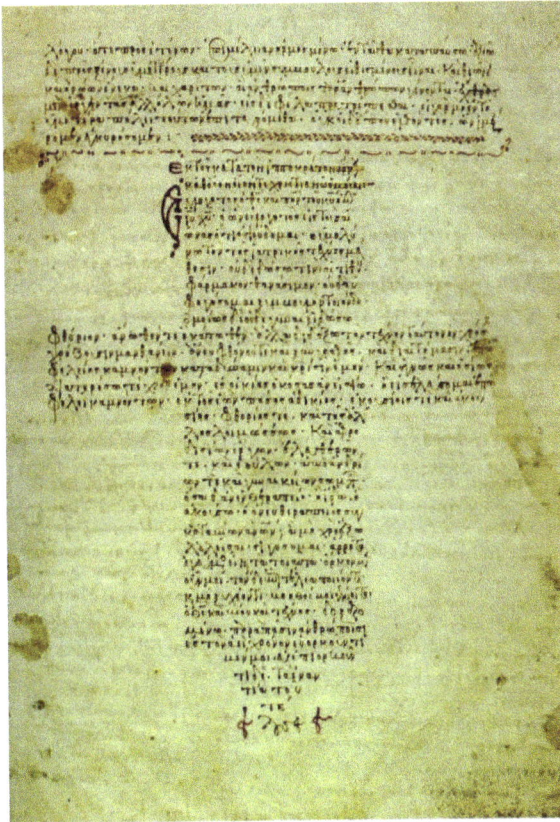

Hippocratic Oath
(Byzantine 1100's).

As previously argued, the distinction between pure physics (trying to understand nature) and applied physics (engineering) is, in practice, artificial. Correspondingly, the relationships between pure biology and medicine and agriculture are complex; ideas flow in both directions.

A central question in the history of science is the difference between the pure and the applied. The meaning of "applied science" is generally agreed on. In contrast, purity implies a desire to understand the natural world with no thought of application. But who's to say that a device or procedure didn't lurk in some dark corner of the scientist's mind? The development of biology, especially reductionist but synthetic as well, has been and still is very closely coupled to medicine. We are ultimately more interested in our mortality than our devices.

The development of medicine is, in part, reflected in the development of anatomy, physiology, embryology, etc. However, on occasion medicine has enjoyed a life of its own, sometimes quite distinct from its contemporary biology.

Hippocrates recognized his very limited ability to heal the sick. He gave comfort to the stricken and provided the family some prognosis so they might plan. He is best remembered for his "oath," which is expected of those who would present themselves as healers. Excerpts and alternate versions are still taught, or at least recited, at graduation from many medical schools. Perhaps the most frequently quoted, with good reason, is "Do no harm."

Most of the initial practices of medicine, East or West, involved lifestyles, herbals, or surgical procedures. Subsequent advances, especially in Europe, are best described, in the following chapters, in terms in advances in the relevant discipline in biology, or in physics. A few are better ascribed to changes in societal norms, especially hygiene and diet. As often quipped, the flush toilet has saved far more lives than has penicillin.

Many of the themes of modern medicine were established by the early 1900s. The development of germ theory rationalized the value of hygiene and sterile techniques. These, in conjunction with the use of anesthetics, transformed surgery. Starting with vaccination against smallpox, inoculations against many diseases were developed. The attendant immune response was first characterized, then analyzed. Antibiotics, both of plant and of synthetic origin, were the first of a broad range of chemotherapeutic drugs. Imaging began with the discovery of X-rays. In parallel, "alternative" forms of medicine were and still are explored. These advances are described in terms of the leading people involved; this perspective admittedly overlooks the contributions of myriad caregivers, clinicians, and researchers.

Early Chinese practices (Chapter A3) provide interesting parallels and contrasts with the intellectual developments of the Mediterranean world over two millennia ago. The Yellow Emperor's Classic of Internal Medicine, *Huangdi Neijing*, was compiled in 2650 B.C. Texts from the reign of Imhotep (3rd dynasty, ~2600 B.C.) in Egypt, described the diagnosis and treatment of 200 diseases.

Perhaps the most important of the teachings of Hippocrates (~460 – ~370 B.C.) was that diseases have natural causes. Quite rightly one explores the psychological, or spiritual, aspects of health and disease, especially in holistic medicine. However, this is different from blaming a demon or invoking a deity.

Dioscorides (~40 – ~90) compiled the first known pharmacopeia, *De materia medica*, about 60 A.D. Zhang Zhongjing (~150 – ~220) wrote the oldest complete medical text, *Shang Han Lun* (*On Cold Diseases*, ~210); it included information on diagnosis, treatment, and prognosis. Medicine was valued and respected in the world of Islam. Avicenna (980–1037) wrote *The Book of Healing* and *The Canon of Medicine* (~1010).

The Mongols and Chinese inoculated people with "attenuated" smallpox virus from dried scabs in the Middle Ages; Pylarini made the first smallpox inoculations in Europe around 1701. It was a century later that Edward Jenner (1749–1823) published *An Inquiry into the Causes & Effects of Variolæ Vaccinæ* (1798), describing his variolations with cowpox (vacca, cow), *Variolæ vaccinæ*.

Oliver Wendell Holmes in 1843 described *The Contagiousness of Puerperal Fever*. Ignaz Semmelweis (1818–1865) knew that this often fatal infection, usually caused by *Streptococcus pyrogenes*, occurred much more frequently in hospital, as opposed to home, deliveries. He demonstrated in 1847 that the incidence could be drastically reduced if doctors simply washed their hands and practiced basic hygiene, a "best practice" that needs to be relearned all too often, in *Die ätiologie, der begriff und die prophylaxis des kindbettfiebers* (*Etiology, Concept and Prophylaxis of Childbed Fever*, 1861). For his trouble, and his Jewishness, he was basically hounded out of the profession and died in an insane asylum in 1865. *Antiseptic Principle of the Practice of Surgery* (1867) by Joseph Lister (1827–1912) was better received and marked a turning point in the practice of medicine.

James Lind (1716–1794) demonstrated in 1747 that citrus fruit, a source of vitamin C, prevents scurvy, a common affliction of sailors on long voyages. The ships of the British navy included lemons or limes with the grub; hence, the slang "limey" for a British sailor. William Withering (1741–1799) described the use of what is now known as digitalis to treat cardiac dropsy, or backward heart failure, in *An Account of the Foxglove* (1785).

Claudius Aymand (1681–1740) in 1735 performed the first successful appendectomy; with no anesthetic, it could not have been pleasant. Humphrey Davy (1778–1829) described the anesthetic properties of laughing gas, nitrous oxide (N_2O), in 1800; however, it was not used clinically until 1844 by an American dentist, Horace Wells. James Blundell (1791–1878) performed the first successful human blood transfusion in 1818; in 1901,

Karl Landsteiner (1868–1943) characterized the different human blood types. Not all innovations in medicine reflected advances in biology. Samuel Hahnemann (1755–1843) founded homeopathy in 1810, as described in *Organon of the Medical Art.*

Claude Bernard (1813–1878) championed a rational approach to understanding and improving the practice of medicine in his *Introduction à l'étude de la médecine expérimentale* (*Introduction to the Study of Experimental Medicine*) in 1865; this text was used well into the 1900s. Robert Koch (1843–1910), along with Pasteur, is credited with articulating the germ theory of disease in 1870. Louis Pasteur (1822–1895) developed vaccines for anthrax in 1881 and for rabies in 1882; it is most fitting that the institute in Paris bears his name.

Wilhelm Röntgen (1845–1923) discovered X-rays in 1895; they were immediately applied to medical imaging. Ivan Pavlov (1849–1936) described the conditional reflex in 1890.

Paul Ehrlich (1854–1915) developed the concept of chemotherapy; his first, of several, successful antibiotics was salarsan (arsphenamine) in 1907, the first effective treatment for syphilis. Andrew Fleming noticed a zone of bacterial clearing around a colony of *Penicillium notatum* in 1929 — a fungal source of antibiotics.

Just as the concept of infectious diseases paralleled the development of cell theory (Chapter C4), so the concept of inborn errors followed advances in genetics (Chapter C11). Archibald Garrod (1857–1936) published *The Incidence of Alkaptonuria: A Study in Chemical Individuality* in 1902 and *Inborn Errors of Metabolism* in 1923. In 1921, Banting and Best purified the hormone, insulin, and used it to treat type I diabetes.

As will be elaborated on in the following chapters, advances in "pure" biology were very frequently associated with or inspired by challenges in medicine.

C2

Anatomy

Andreas Vesalius
(1514—·1564).

The investigations of many disciplines, such as astronomy, geology, chemistry, and paleontology, begin by describing "what is there" as accessible to human sensibilities, primarily but not exclusively vision, given the tools at hand, e.g. a knife or a simple lens. Gross anatomy was known to early hunters, farmers, and artists. Aristotle recorded the external morphology and internal anatomy of numerous vertebrates and invertebrates.

By convention "anatomy" refers to human anatomy and contains the implicit assumption that, at some levels, humans are similar to other mammals. Access to human cadavers for dissection varied widely with cultural and legal norms. Galen was not the only anatomist of the Roman Empire, but his works, written in Greek, though he lived much of his life in Rome, were translated to Arabic and regarded in the Arab world as a standard reference. Some details were challenged by physicians like Avicenna. Galen was subsequently translated to Latin and treated as gospel by the early anatomists in several universities of northern Italy and Paris. The questioning of Galen and the publication in 1543 of *De humani corporis fabrica* by Vesalius was a turning point in the study of anatomy. Although not so important as Copernicus' *Revolutionibus*, also published in 1543, it is one of the seminal events of the Renaissance, and 1543 is sometimes cited as the beginning of the scientific revolution. The Church was not threatened; there followed a parade of great anatomists, many at Padua, exploring the human body in ever greater detail. Inevitably, they speculated about function and development.

Aristotle (384–322 B.C.) recorded the external morphology and internal anatomy of numerous vertebrates and invertebrates. Several distinguished Greek anatomists preceded Galen. Herophilos (335–280 B.C.) and his younger colleague, Erasistratus (304–250 B.C.), established a school of anatomy in Alexandria where they performed the first human dissections in the Greek world. They argued, as quoted by Galen, that the brain is the center of the "nervous system" and the site of intelligence. They distinguished blood vessels from nerves and further distinguished motor from sensory nerves. They noted valves of the heart, recorded palpitations, and stated that the heart functions as a pump.

Galen (129–~208) was a surgeon in the gladiator school in Pergamum, where he coined the evocative phrase, "windows into the body." He extended his dissections to macaques (Barbary apes). He distinguished light arterial blood from dark blood from the liver and endorsed "bloodletting" to reduce the latter — a practice favored by followers of Asclepius for over 1,500 years. Later, he served as physician to Emperors Marcus Aurelius and Septimius Severus in Rome. Some details of his writings were challenged by physicians like Avicenna (980–1037).

Ibn al-Nafis (1213–1288) was chief of physicians at the Al-Mansouri hospital in Cairo, where he wrote the *Commentary on the Anatomy of Canon of Avicenna* (1242). He denied the presence of pores through the inter-ventricular septum and stated that blood from the right ventricle goes to the lungs, where the lighter parts filter into the pulmonary vein, mix with air, and return to the left ventricle. The ventricle is nourished from blood in vessels in its own substance. "The lungs are composed of parts, one of which is the bronchi; the second, the branches of the arteria venosa; and the third, the branches of the vena arteriosa, all of them connected by loose porous flesh." "The permeation of arteries into the cranium is well known not to be from the front ventricle. Pulsation results from heart contractions; he identified ten cranial nerves "... each nerve (of the eye) goes to the opposite side." "The most important muscles of a human body total 529 ..."

Leonardo da Vinci (1452–1519) participated in many dissections, most at the hospital Santa Maria Nuova, Florence. He noted that bronchi of the lung have a blind end and that under pressure no air is transmitted to the left auricle. He described four cavities of the heart and noted that ventricular valves prevent regurgitation. He recognized the difference in maternal and embryonic circulations. He and other artists, most notably Masaccio, painted figures in the round and were concerned with the full dimension of man, not just his two dimensional projection. Bartolomeo Eustachi (~1500–1574), a contemporary of Vesalius, completed a lovely set of anatomical drawings in 1552, first bound and published as *Tabulae anatomicae* (1714). He described in detail the internal and anterior muscles of the malleus and the stapedius as well as the complex cochlea and, of course, the Eustachian tube. He described the adrenal glands in 1563.

Andreas Vesalius (1514–1564) was born in Brussels to a family of physicians. He studied at Pavia, Leuven, then Paris, and became chair of surgery at the University of Padua in 1537. He had access to executed criminals and, contrary to the practice of the time, performed his own dissections before students in an amphitheater that well merits a visit today.

> When I undertake the dissection of a human cadaver I pass a stout rope tied like a noose beneath the lower jaw and through the zygomas up to the top of the head ... The lower end of the noose I run through a pulley fixed to a beam in the room so that I may raise or lower the cadaver as it hangs there or turn around in any direction to suit my purpose ... You must take care not to put the noose around the neck, unless some of the muscles connected to the occipital bone have already been cut away.

His publication of *De humani corporis fabrica* in 1543 is rightly regarded as a major event in biology. In it he corrected many of Galen's errors: Blood does not pass through the ventricular septum.

The sphenoid bone is well described. The sternum consists of three segments and the sacrum of five vertebrae. He verified the valves of the hepatic veins and described the umbilical vein and its course to the vena cava and ductus venosus. He correctly described the spleen, colon, and pylorus. He gave the first account of the mediastinum, pleura, and caecal appendix as well as the connections of the greater omentum to the stomach.

Matteo Renaldo Colombo (~1516–1559) succeeded Vesalius as professor of anatomy and surgery at Padua (1544–1559). In *De re anatomica* (1559), he wrote:

> Between the ventricles is the septum, through which almost all think there is a way from the right ventricle to the left, so that the blood in transit may be rendered subtle by generation of the vital spirits in order that the passage may take place more easily. This, however, is an error; for the blood is carried by the arterial vein [pulmonary artery] to the lung ... It is brought back thence together with air by the venal artery [pulmonary vein] to the left ventricle of the heart. This fact no one has hitherto observed or recorded in writing; yet it may be most readily seen by any one ... Wherefore I cannot wonder enough that anatomists have not observed a matter so clear and of such importance. For them it suffices that Galen said so. There are some in our time who swear by the opinions of Galen and assert that he should be taken as gospel, and that there is nothing untrue in his writings.

It is important to get this anatomy right as the foundation for future studies of function, that is, physiology. Of greater importance and consistent with the spirit of the Renaissance, the teachings of the Greeks were questioned and subject to experimentation.

Gabriele Falloppio (1523–1562) was chair of anatomy at the University of Pisa (1548–1551) and at Padua (1551–1562). He was prolific — *Observationes anatomicae* (1561); *De partibus similaribus humani corporis*; *Opera genuina omnia*; *Expositio in librum Galeni de ossibus*; *Observationes de venis*; *De humani corporis anatome compendium* (1571). He described primary dentition, growth of the tooth bud and growth and replacement by the secondary tooth. He lent his name to the Fallopian tubes, which bear eggs to the uterus. He detailed the ureter and renal vessels, three muscle coats of the urinary bladder, the valvulae conniventes of the small intestine. He described the clitoris and hymen and disproved the commonplace that the penis enters uterus. His engineering application consisted of designing condoms. "I tried the experiment (use of condoms) on 1,100 men and I call immortal God to witness that not one of them was infected." (with *Treponema pallidum*, syphilis).

Hieronymus Fabricius (1537–1619) followed Falloppio as chair of anatomy and surgery in 1562. He described the early anatomy of the fetus and the valves in veins. "In my opinion these valves are formed that they may to a certain extent delay the blood and so prevent the whole of it flowing to the feet, the hands, or the fingers and collecting there."

> In respiration, Nature sets herself a double end, the generation of animal spirits and the regulation and maintenance of the innate heat. The heart is regulated by [a relation between] the fuel supplied, refrigeration [by the lung], and the elimination of the superfluities. All these are the result of the air taken into the body, whence the necessity for respiration … Respiration is the movement of air by which spirit is taken in and given out through the mouth. In inspiration air enters the lung and the heart, carrying material and coldness; in expiration on the other hand the superfluous residues are evacuated.

Understanding anatomy does not guarantee a full understanding of function but it sets one thinking; it gives one a model from which to pose new questions.

As will be discussed in Chapter C10, the study of human anatomy was not the only focus of biology during the Renaissance. Conrad Gessner (1516–1565) extended his interests in botany and linguistics to zoology and comparative anatomy, *Historiae animalium*, (5 Vols., 1551–1558).

More discoveries would be made, but by the beginning of the scientific revolution basic human anatomy was understood. How does it work? *Vide infra* (Chapter C3).

C3

Physiology

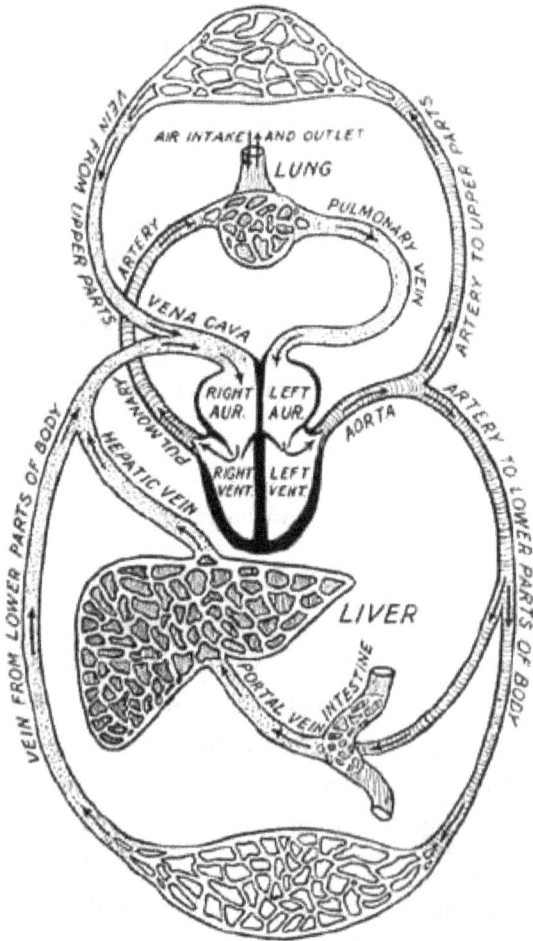

Circulation of blood.

Certainly the Greeks, the Chinese, and the Muslims speculated about how the body works. Some educated guesses were sort of correct; others were wildly fanciful. One of the results of the anatomical studies of the 1500s in Europe was that physicians and anatomists began to pose more focused and nuanced questions about mechanisms (physiology) and development (embryology). The refinement and extension of these questions, and answers, can be traced to contemporary research.

William Harvey published *Exercitatio anatomica de motu cordis et sanguinis in animalibus* in 1628 (dedicated to Charles I), 59 years before Newton's *Principia* and only two years after the death of Francis Bacon. His postulated mechanism of the circulation of blood rested on a detailed understanding of anatomy, especially the location and function of valves in the heart and veins. He applied quantitative analysis, i.e. comparing cardiac output per time with total blood volume, and he made the bold leap of postulating a return flow of arterial blood to the venous system via yet-to-be-identified capillary beds.

His work set the highest standard: choice of a significant and tractable problem, familiarity with the studies of others, hands on experience, choice of appropriate parameters, quantitative analysis, and formulation of a model that drove additional research. His was one of the landmark discoveries of the scientific revolution.

Galen (129–~208) had noted that blood from the liver is dark and that arterial blood is light, a fringe benefit of working with gladiators. The innate heat, generated by the heart, is cooled by air from the lungs brought to the heart by the pulmonary vein. He referred to "blood pulsing back and forth like tide." Given the anatomy, this was a reasonable hypothesis.

Ibn al-Nafis (1213–1288) denied that there are pores in the interventricular septum, as stated by Galen, and rightly proposed that blood from the right ventricle goes to the lungs. He then postulated that the lighter parts of the blood filter into the pulmonary vein, mix with air, and return to the left ventricle — correct, with a bit of leeway on "mix."

> The primary purpose of the expansion and contraction of the heart is to absorb the cool air and expel the wastes of the spirit and the warm air; however, the ventricle of the heart is wide. Moreover, when it expands it is not possible for it to absorb air until it is full, for that would then ruin the temperament of the spirit, its substance and texture, as well as the temperament of the heart. Thus, the heart is necessarily forced to complete its fill by absorbing the spirit.
> —*Commentary on the anatomy of the canon of Avicenna* (1242).

This is a neat example of the transition from spirits, however defined, to logical analysis.

Leonardo da Vinci (1452–1519) noted that the ventricular valves prevent regurgitation, an essential component of the circulatory system.

Miguel Serveto Conesa (1511–1553, see Chapter A9) elaborated that "... the vital spirit passes from the arteries to the veins through the anastomoses."

> The vital spirit is generated through the mingling in the lungs of the inspired air with the subtle blood which is communicated to it from the right ventricle to the left. This communication does not, however, take place, as is generally believed, through the

septum of the heart, by a remarkable device the subtle blood is driven from the right ventricle through a long passage in the lungs. It is prepared by the lungs, and is there rendered lighter in color, and from the artery like vein [pulmonary vein] it is mixed with the inspired air, and by expiration is cleansed from its fumes. So at length completely mingled [with the air] it is drawn in by the left ventricle during its expansion, ready to become vital spirit ...

That communication and preparation ... does not take place in this way through the lungs [and not through the septum] is moreover shown by the manifold conjunction and communication of the arterial vein [pulmonary artery] and the venal artery [pulmonary vein] in the substance of the lung. This is confirmed by the remarkable size of the venal artery [pulmonary vein], which would not have been made so large and would not discharge from the heart itself such a mass of pure blood into the lungs merely for nourishment.

All but three copies of Servetus' manuscript, *Christianismi restitutio* (1563), had been destroyed; Harvey rediscovered them.

Matteo Realdo Colombo (~1516–1559) discussed the pulmonary circuit:

Between the ventricles is the septum, through which almost all think there is a way from the right ventricle to the left, so that the blood in transit may be rendered subtle by generation of the vital spirits in order that the passage may take place more easily. This, however, is an error; for the blood is carried by the arterial vein [pulmonary artery] to the lung ...It is brought back thence together with air by the venal artery [pulmonary vein] to the left ventricle of the heart.

Hieronymus Fabricius (1537–1619) wrote in 1574: "In my opinion these valves are formed that they may to a certain extent delay the blood and so prevent the whole of it flowing to the feet, the hands, or the fingers and collecting there." Aristotle had written that "... the heart burns with a flame." This was still accepted.

In respiration, Nature sets herself a double end, the generation of animal spirits and the regulation and maintenance of the innate heat. The heart is regulated by [a relation between] the fuel supplied, refrigeration [by the lung], and the elimination of the superfluities. All these are the result of the air take into the body, whence the necessity for respiration … Respiration is the movement of air by which spirit is taken in and given out through the mouth. In inspiration air enters the lung and the heart, carrying material and coldness; in expiration on the other hand the superfluous residues are evacuated.

William Harvey (1578–1657) received his B.A. from Gonville and Caius College, Cambridge, in 1597. He studied under Fabricius in Padua in 1602 and practiced medicine at St Bartholomew's Hospital (1609–1643). He was personal physician to King James I (1618–1625) and to Charles I (1625–1647). He unraveled the circulatory system in 1616 and published *Exercitatio anatomica de motu cordis et sanguinis in animalibus* (*An Anatomical Observation Concerning the Motion of the Heart and Blood in Animals*), 71 years before Newton's *Principia* (Chapter B4). His reasoning is contemporary; his exposition lovely (as condensed):

1) "If the heart is grasped in the hand, it may be felt to become harder during its action. This hardness proceeds from tension, just as when the forearm is grasped, its muscles are perceived to become tense and firm when [these muscles] move the fingers."
2) "It may further be observed in fishes and cold blooded animals [in which the heart beats long after death] … that when the heart moves it becomes a lighter color, and when it is quiescent it becomes of a deeper blood-red color."
3) Arteries expand immediately following contraction of the heart producing a pulse.
4) Blood gushes from a cut artery in spurts.
5) In the dying heart the ventricles cease beating before the auricles. If the ventricles are exposed, one can see blood enter from the auricles with each beat.
6) In the intact heart the auricles contract before the ventricles.
7) Once blood has entered the aorta, or the pulmonary artery, it cannot return (bicuspid, tricuspid valves).
8) 72 beats/minutes $\cdot$ 60 minutes/hour $\cdot$ 2 ounces/beat = 8640 oz/hr.; the blood must circulate.

9) One can bleed to death from a cut vein; hence, communication between artery and vein.
10) Blood from right side of the must pass through lungs.
11) "I ... tremble lest I have mankind at large for my opponents ... Doctrine, once sown, strikes its root deep, and respect for antiquity influences all men ... I had long considered what was the quantity of blood transmitted, and in how short a time its passage might be effected. I did not find it possible that it could be supplied by the juices of the ingested ... unless the blood should somehow find its way from the arteries into the veins, and so return to the right side of the heart." "I now began to ask whether there might not be a movement, as it were, in a circle, and this I afterwards found to be true. I saw that the blood, forced by the action of the left ventricle into the arteries, was distributed to the body at large, and its several parts. In the same manner it is sent through the lungs, impelled by the right ventricle into the arterial vein [pulmonary artery]."
12) "And now the cause is manifest, why in our dissections we usually find so large a quantity of blood in the veins, so little in the arteries; why there is much in the right ventricle, so little in the left, which probably led the ancients to believe that the arteries contained nothing but spirits during the life of an animal. The true cause of the difference is this, that there is not passage to the arteries, save through the lungs and heart. When an animal has ceased to breathe and the lungs to move, the blood in the arterial vein [pulmonary artery] is prevented from passing into the venal artery [pulmonary vein] and from thence into the left ventricle of the heart ... But the heart not ceasing to act the same precise moment as the lungs, but surviving them and continuing to pulsate for a time, the left ventricle and the arteries go on distributing their blood to the body at large and sending it into the veins; receiving none from the lungs, however, they are soon exhausted, and

left empty." "We are now in a condition to suspect why no one has yet said anything to the purpose upon the anastomosis of veins and arteries, either whether or how it is affected or for what purpose. I now enter on an investigation of that subject."

13) The reptilian heart continues beating after death; clamp the vena cava and the ventricle empties. "But if the artery instead of the vein is compressed, you will observe the part between the obstacle and the heart, and heart itself, to become inordinately distended, to assume a deep purple or even livid color. At length it is so much oppressed with blood that you will it about to be choked. When, however, the obstacle is removed, all things immediately return to their natural state in color, size, and impulse."

14) "Now make an experiment on the arm of a man, using a bandage as employed in blood-letting. The best subject is a lean man who has large veins ... Let the bandage be tied round the arm and drawn as tightly as can be borne. It will first be perceived that beyond the bandage, neither in the wrist nor elsewhere, do the arteries pulsate. Immediately above the bandage, however, the artery rises higher at each expansion and throbs more violently ... as if it strove to break through the obstacle to its current ... The hand retains its natural color and appearance, though in the course of time it begins to get somewhat colder." "After the bandage has been kept thus for some time, loosen it a little. The whole hand and arm will now instantly become deeply colored and distended, and the veins themselves tumid and knotted. After ten or twelve pulses the hand will be excessively distended, injected, gorged with blood. Now apply the finger attentively over the artery which is pulsating near the bandage. At the moment of slackening the blood will be felt to glide underneath the finger. Moreover, he on whose arm the experiment is being made is distinctly conscious of a sensation of warmth as the bandage is slackened. He feels too something,

namely, a stream of blood, suddenly making its way along the course of the vessels and diffusing itself through the hand, which at the same time begins to feel hot, and becomes distended." "In connection with the tight bandage we noted that the artery was distended and pulsated above the bandage, but not below it. In the case of the moderately tight bandage, on the contrary, we find the veins below, never above, the bandage, swell, and become dilated, while the arteries [below it] shrink..." "The moderately tight bandage then renders the veins turgid and distended, and the hand full of blood. I ask, therefore, whence is this? Does the blood that accumulates below the bandage come through the veins, or through the arteries, or does it pass by certain hidden pores? Through the veins it cannot come. Still less can it come through any system of invisible pores. It must needs then arrive by the arteries, in conformity with all that has been already said. That it cannot flow in by the veins appears plain from the fact that the blood cannot be forced towards the heart unless the bandage be removed, and when this is done, suddenly all the veins collapse, and disgorge themselves of their contents into the superior parts, the hand at the same time resuming its natural pale color. Moreover, if a man has had his arm thus bound for some little time with a moderately tight bandage, so that it has not only got swollen and livid, but cold, and if the bandage then is loosened, he feels something cold make its way upwards along with the returning blood..." "Further, when the bandage is relaxed from extreme tightness [to moderate tightness], we see the veins below the bandage instantly swell up and become gorged, while the arteries meantime continue unaffected, this is an obvious indication that the blood passes from the arteries into the veins, and not from the veins into the arteries." "This shows that there must therefore be either an anastomosis of the two [kinds] of vessels, or passages in the flesh and solid parts that arc permeable by the blood. It shows too that the veins themselves have frequent communications with one another, because they all become turgid together, with the moderately tight bandage and, moreover, if any single small vein be then pricked with a lancet, they all speedily shrink."

15) "We have now spoken of the blood that passes through the heart and the lungs [and of the blood that passes] from the arteries into the veins in the peripheral parts and the body at large. We have yet to explain, however, in what manner the blood finds its way back to the heart from the extremities by the veins, and how and in what way these are the only vessels that convey the blood from the external to the central parts." "The celebrated Hieronymus Fabricius of Aquapendente, a most skilful Anatomist, and venerable old man made representations of valves in the veins. These are situated at different distances from each other, and diversely in different individuals ... and are directed upwards or towards the trunks of the veins. If therefore anything attempted to pass from the trunks into the branches of the veins, or from the greater vessels into the less, they completely prevent it..." "The discoverer of these valves did not rightly understand their use, nor have others added anything to our knowledge. Their function is by no means explained when we are told that it is to hinder the blood, by its weight, from all flowing into inferior parts." They do not, in fact, always do this, "for the valves in the jugular veins [in the neck] hang downwards, and are so arranged that they prevent the blood from rising upwards; in a word, the valves do not invariably look upwards, but always towards the trunks of the veins, invariably towards the seat of the heart."

16) "These valves are solely lest the blood should pass from the greater into the lesser veins, lest, instead of advancing from the extreme to the central parts of the body, the blood should rather proceed along the veins from the center to the extremities." "And this I have frequently experienced in dissections

of the veins. If I attempted to pass a probe from the trunk of the veins into one of the smaller branches, I found it impossible to introduce it far by reason of the valves; whilst, on the contrary, it was easy to push it along in the opposite direction, from the branches towards the trunk."

"That all this may be made the more apparent, let an arm be tied up above the elbow at A. In the course of the veins, especially in laboring men and those whose veins are large, are certain knots or elevations as at B, C, D, E, and F, which will now be seen. These knots are not only at the places where the veins branch, as at E and F, but also where they do not, as at C and D. These knots are formed by valves, which thus show themselves externally." "If you now press blood from the space above one of the valves, as from H to O, and keep the point of the finger upon the vein below, you will see no influx of blood from above. The portion of the vein between the point of the finger and the valve O will remain empty. Yet the vessel will continue sufficiently distended above that valve, as at O, G. If you now apply a finger of the other hand upon the distended part of the vein above the valve O, and press downwards, you will find that you cannot force the blood through or beyond the valve. You will only see the portion of vein between the finger and the valve becomes more distended, while the portion of the vein below the valve (H, O) still remains empty." "It therefore appears that the function of the valves in the veins is the same as that of the three sigmoid valves at the commencement of the aorta and pulmonary artery, namely, to prevent all reflux of the blood that is passing through them." "Further, the arm being bound at A, A as before, and the veins full and distended, compress a vein with one finger L, at a point below a valve or knot. Then with another finger stroke the blood upwards beyond the next valve N. You will now perceive that this portion of the vein L, N still continues empty. But if the finger first applied (H, L) is removed, the vein is immediately filled from below, and the arm becomes again as at D, C." "That the blood in the veins therefore proceeds from more remote to less remote parts and towards the heart, moving always in the vessels in this, and not in the contrary direction, appears most plain for the veins are free and open conduits of the blood returning to the heart, and are yet effectually prevented from serving as its channels of distribution from the heart."

17) "Now I may give my view of the circulation of the blood and propose it for general adoption." "All things, both argument and ocular demonstration, confirm that the blood passes through lungs and heart by the force of the ventricles, and is driven thence and sent forth to all parts of the body. There it makes its way into the veins and pores of the flesh. It flows by the veins everywhere from the circumference to the center, from the lesser to the greater veins, and by them is discharged into the vena cava and finally into the right auricle of the heart. The blood is sent in such a quantity, in one direction, by the arteries, in the other direction by the veins, as cannot possibly be supplied by the ingested food. It is therefore necessary to conclude that the blood in the animals is impelled in a circle, and is in a state of ceaseless movement; that this is the act or function of the heart, which it performs by means of its pulse; and that it is the sole and only end of the movement and pulse of the heart."

Harvey had to propose an unknown, microscopic capillary bed connecting arterial and venous circulations. His observations, analyses, and articulation could well serve as a model for contemporary physiologists in particular, biologists in general. This in 1616, at the time of Francis Bacon. Harvey's *Motu cordis* set a high standard for subsequent studies. For instance Alfonso Borelli (1608–1679) published a sophisticated, mechanical analysis, *On the Movement of Animals* (1680).

Marcello Malpighi (1628–1694) had a broad range of interests; he was the first to describe taste buds and fingerprints. Most relevantly, he described the capillary beds of the lung in *Anatomical Observations on the Lungs* (of the frog), presented to the Royal Society in 1661. "As the blood streams, thus repeatedly divided up, is carried round in a sinuous manner, its color fades. It is thus distributed until it approaches the walls ... receiving branches of the veins ... While the heart is still beating, two movements in opposite directions can be seen, making the circulation of the blood quite evident." Subsequent publications included *De polypo cordis* (blood clotting, 1666) and *Anatomia plantarum* (plant structure, 1671).

Antonie van Leeuwenhoek (1632–1723) observed blood flowing in capillaries in the webs of frogs' feet in 1674 and regarded this as evidence for circulation. Jan van Swammerdam (1637–1680) wrote: "I saw a serum in the [frog] blood in which were a vast number of roundish particles, of a flat, oval, but regular form ... When I viewed them sideways they resembled crystalline clubs, or other figures, according as they were turned about in various directions in the serum of the blood," *Bybel der natuur (1668)*. *He removed the heart from a frog and could still induce its leg to twitch by irritating its motor nerve or its brain; "... motion or irritation of the nerve alone is necessary to produce muscular motion" (Chapter C9).*

Stephen Hales (1677–1761) wrote Vegetable staticks (1727, Chapter C10) and Haemastaticks (1737), in which he measured the capacities of the heart and of different vessels, the "force of the blood," and the flow rate in different animals. Thomas Jefferson opined: "Harvey's discovery of the circulation of the blood was a beautiful addition to our knowledge of the animal economy, but on a review of the practice of medicine before and since that epoch, I do not see any great amelioration which has been derived from that discovery." He was too impatient. René-Théophile-Hyacinthe Laennec (1781–1826) invented the stethoscope while working in the Hôpital Necker, De l'auscultation médiate ou traité du diagnostic des maladies des poumons et du coeur (1819). Today, translational research is judged in months, not centuries.

C4

Cell Biology

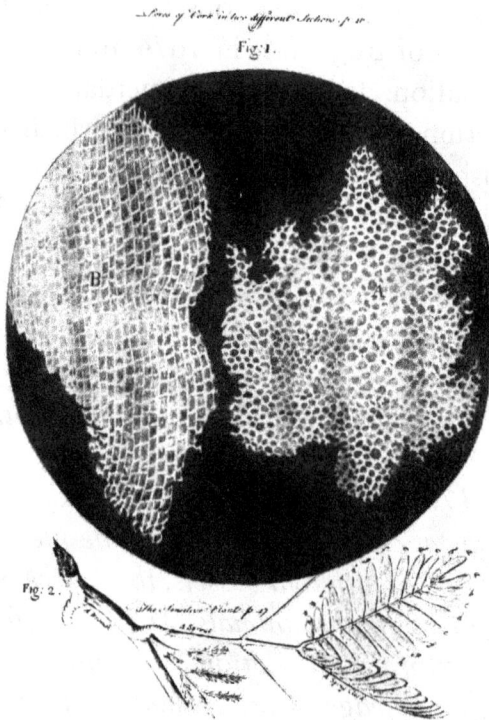

Micrographia, 1665
(Robert Hooke).

The names given to various disciplines of science are artificial, both in content and in dates. Creative scientists, engineers, and physicians frequently, knowingly or not, borrow ideas and implements from other practitioners. The division of this book into distinct chapters is a necessary literary device to facilitate telling and reading the story; however, it creates an artificial sense of partition. Apologies.

The extensions of the study of anatomy include organismal physiology, comparative physiology, and cytology — all still active areas of research. Most important, anatomists, not unlike their fellow chemists, began to seek the fundamental unit(s) of these organs. As is so often the case, the field could pose new questions because of a technical advance, the lens. Just as new insights in science lead to unanticipated applications, so too new tools benefit fields of research not envisioned by their smithies. Hooke, using a simple lens, observed regular outlines in the bark of cork and called them cellules after the tightly packed cells of monks. Leeuwenhoek improved the lenses and extended these observations to many other types of tissues.

The journey to the "cell theory" (1738) of Schleiden and Schwann covered two centuries. The origin(s) of cells, their replication, their nuclei and other organelles, their differentiation, and their interactions are still the subjects of intense and rewarding research.

Robert Hooke (1635–1703), looking through his 3x lens, saw regular partitions in cork; he called them *cellula* after the rooms of monks, *Micrographia* (1665). Antonie van Leeuwenhoek (1632–1723) was an apprentice with a Scottish cloth merchant in Amsterdam. He used a lens to examine the fibers of cloth. His lens grinding ability developed to 275x, and his curiosity to striated muscle fibers, blood flow in capillaries, spermatozoa, and "animalcules," as *giardia*. As noted in the preceding chapter, his study of the capillaries in the web of frogs' feet helped complete the circuit for Harvey. Jan Swammerdam (1637–1680), also from Amsterdam, observed red blood cells in 1668 as subsequently described in *Bybel der natuure*. Lazzaro Spallanzani (1729–1799) was on the faculties of the Universities of Bologna, Reggio, Modena, and Pavia. He pondered spontaneous generation and suggested that microbes travel through the air and can be killed by boiling (1768).

Matthias Jakob Schleiden (1804–1881) studied law at the University of Heidelberg; however, his studies of plants led to professorships at Jena, then Dorpat (1863). He argued that plants are composed of cells, as are animals. He observed that the cell nucleus divides as the cell divides. He leaned toward vitalism and suggested that cells are ordered by the same forces that obtain in crystals. Theodor Schwann (1810–1881) held professorships at the Universities of Leeuven and of Lüttich (1848). He maintained that fermentation is a biological, as opposed to a chemical, process and results from an airborne factor. He dined with Schleiden in Berlin in 1837, and continued the conversation by mail. Together they formulated two ideas of what came to be called the "cell theory."

All organisms are composed of one or more cells. The cell is the basic unit of structure, function, and organization in all organisms. However, neither Schleiden nor Schwann supported the subsequent postulate of Virchow, *Omnis cellula e cellula*.

Rudolf Ludwig Karl Virchow (1821–1902) graduated from the military academy in Berlin in 1842 then attended medical school. He argued "Every cell from a cell" in 1858. He is remembered in the clinical literature for his work in cellular pathology and specifically, for his "triad" describing a pulmonary thrombo-embolism: Formation of a venous thrombus. It passes through the heart to the pulmonary artery; and lodges in the lungs (unless it finds a by-pass via patent ductus arteriosis).

Louis Pasteur (1822–1895) contributed to many fields. His studies of contamination in general and especially fermentation led to the conclusion that the causative agents, microbes, could be airborne. These insights in turn led to the technique of pasteurization of milk and wine. He endorsed the antiseptic surgery of Lister.

Sydney Ringer (1836–1910) completed his M.D. at University College, London, in 1863; his *Handbook of Therapeutics* went through 13 editions (1869–1897). He researched the characteristics of various organs, especially the heart, as dissected and maintained beating *in vitro*. His assistant used water supplied by the "New River" company, instead of their usual distilled water; this lead to the discovery that calcium is required in the bathing solution for organs and tissues and to the formulation of the "Ringer solution."

Walther Flemming (1843–1905) was a military physician in the Franco-Prussian war, then Professor of Anatomy at the University of Kiel (1876–1901). He stained nuclei with aniline dyes and described mitosis in detail. He proposed *omnis nucleus e nucleo*, after Virchow's *omnis cellula e cellula*.

Robert Koch (1843–1910) developed techniques — sterilization by boiling, use of "agar" from Japanese seaweed, *Agar agar* — essential to microbiology. He cultured several pathogenic bacteria: *Bacillus anthracis* (1877), *Mycobacterium tuberculosis* (1882), and *Vibrio cholera* (1883). His criteria for attributing a disease to a pathogen (1882) are still honored today — essentially the "germ theory":

1) The organism is found in all cases of the disease examined.

2) It should be prepared and maintained in a pure culture.

3) It is capable of producing the original infection, even after several generations in culture.

4) It should be retrievable from an inoculated animal and cultured again.

Camillo Golgi (1843–1926) extended anatomy to higher resolution, that is, histology. This required better microscopes and, fully as important but perhaps less well appreciated, the development of specific stains. He identified the sensor of tension in tendons (Golgi receptor), distal tubules in the kidney and the intracellular Golgi apparatus, a convoluted extension of the outer membrane of the nucleus (1898). Using silver nitrate stains, he identified the cell bodies of neurons in sections of brain and their extensions into a "reticulum."

Santiago Ramón y Cajal (1852–1934) accompanied the Spanish military expedition to Cuba in 1874. He held several positions — professor at the University of Valencia (1881); director of the Zaragoza Museum (1879); director of the National Institute of Hygiene (1899); and founder of Laboratorio de Investigaciones Biológicas (1922). He is remembered for his histology of the nervous system using, after Golgi, silver chromate. His interpretation was millions of separate cells, neurons with their extended axons, not a reticulum as posited by Golgi. He proposed communication via "electrical junctions," now called synapses. Ramón y Cajal and Golgi shared the Nobel Prize in 1906.

C5

Embryology

Man the seed, woman the incubator

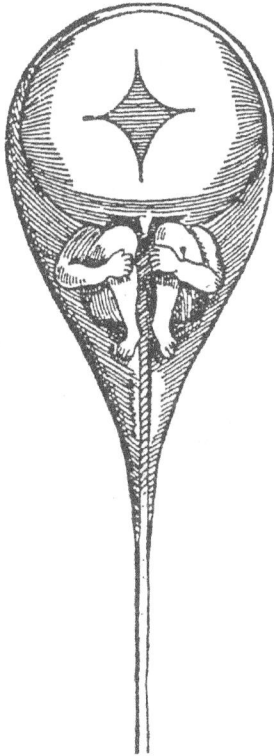

Homunculus
The little pre-formed person in the sperm. An imaginary representation of what a sperm might look like, if able to be seen clearly, drawn by Nicolaus Hartsoeker in *Essai de diotropique*, 1694.

Homunculus (Nicolas Hartsoeker, 1695).

Various societies had a range of myths concerning reproduction and early development, often paralleling those about the first humans and the origin of life (Chapter C17). Hippocrates suggested that the fetus developed from coagulated menstrual blood. Later Greeks speculated as to the contributions of mother and father. Some held that the infant was the product of the mother and that the father only initiated the process; others regarded the mother as an incubator, providing only nutrition. These theories do not reveal an obvious correlation with the roles of women in the societies and with the relative dominance of men. The umbilical cord was recognized as the root to the uterus. The discussions as to when the soul enters the fetus remain relevant today.

The Greeks realized that the embryos of humans and of other vertebrates resemble the adults but were not mere miniatures. The speculation as to whether a homunculus could contain the next generation's homunculus was not really resolved until the 1940s! This fundamental question of how the embryo develops into an adult has been reformulated as ever more is understood of eggs and sperm, of homology, of DNA, of stem cells, and of control of gene expression. Still today, it is one of the fundamental challenges of reductionist biology. How does the information encoded in the sequence of four bases determine the complex structure and behavior of an adult human (or fungus)?

Hippocrates (~460–~370 B.C.) suggested that since there is no menstruation during pregnancy, the fetus develops from coagulated blood. Seen in context, this was a reasonable hypothesis. As embryology developed over the millennia one might ask whether there was a single, or several shift(s) in paradigm.

Aeschylus relates in his play, *Eumenides*, that Apollo defended Orestes, charged with matricide: "The mother ... of the child is no parent of it but nurse only of the young life that is sown in her. He is the male and she but a stranger, a friend, who, if fate spares his plant, preserves it till it puts forth." The "same" plant may be grown from (male) seed sown on different (female) soils. When an enemy was defeated, the males were killed; the females were taken as concubines (incubators) and would not compromise offspring fathered by the victors. This strategy is adopted by many mammals.

Aristotle (384–322 B.C.) noted the mammalian-like embryo of the hound shark, *Mustelus laevis*, and the "stages" of development of the chick embryo in *On the Generation of Animals*. He asked in *De Anima*: "O ye Mothers, and say whether you do not feel the movements of the child with you. How then can it have not soul?" He, and colleagues, described the unfertilized egg as a complex machine, ready to go. He distinguished primary vs. secondary sexual characteristics and realized that the sex can be determined early in the life of the embryo. He appreciated the parallels between regeneration and embryonic development and understood the function of the umbilical cord and the placenta, "root to the womb." However, some interpretations were incorrect. Aristotle felt that semen must be derived from all parts of the body since it forms all parts in the offspring — quite reasonable. He missed the role of the testes. He thought that caterpillars derived from "eggs laid too soon" and did not appreciate the complex sequence of egg, larva (caterpillar), pupa, adult in insects.

Galen (129–~208), in *On the Formation of the Foetus*, referred to birth as cessation of the retentive faculty of the uterus. He appreciated the importance of different tissues but could hardly have guessed the mechanisms of differentiation. "Genesis is not a simple activity of Nature, but is compounded of alteration and of shaping. That is to say, in that bone, nerve, veins and all other tissues may come into existence, the underlying substance from which the animal springs must be altered…shaping or formative process/activity."

Ibn al-Nafis (1213–1288) wrote:

> Galen believes that each of the two semen has in it the active faculty to fashion and the passive faculty to be fashioned; however, the active faculty is stronger in the male semen while passive in the female semen. The investigators amongst the falasifa (followers of the philosopher Al-Ghazali) believe that the male semen only has the active faculty, while the female only has the passive faculty…As for our opinion on this, and God knows best, neither of the two semen has in it an active faculty to fashion.

"… once the male semen and female semen are brought together in the womb, the female semen quenches the hot fire of the male semen through its own cool and wet nature."

Leonardo da Vinci (1452–1519) recognized different embryonic vs. maternal circulations. Further, he stated: "The liver is relatively much larger in the fetus than in the grown man." "… the seed of the female is as potent as that of the male in generation" (Chapter C2). Gabriele Falloppio (1523–1562) was appointed chair of Anatomy and Surgery at Padua in 1551; he recognized the *aquæductus Fallopii* (Fallopian tube). Volcher Coiter (1534–1576) was born in Groningen; he studied with Eustachius and Falloppio and described the follicles and *corpor lutea* of the ovary. He is considered "father of embryology" in the Netherlands. Hieronymus Fabricius (1537–1619) was appointed professor of anatomy and surgery in 1562; he described in more detail the formation of the human foetus and is "father of embryology" in Italy.

William Harvey (1578–1657) endorsed *omne vivum ex ovo* (all life from an egg), but never identified a mammalian egg. Marcello Malpighi (1628–1694) also hypothesized the mammalian egg but failed to find it. Given the size of the eggs of other vertebrates, how could they imagine how miniscule our eggs are? Antony van Leeuwenhoek (1632–1723) described a motile "homunculus," semen. This concept of a preformed "small man" had been rumored for centuries, as far back as Paracelsus. Nicolas Hartsoekker in 1695 made one of many symbolic drawings. They illustrated the fundamental paradox of a homunculus within a homunculus, within, within … for how many generations? Rudolf Ludwig Karl Virchow (1821–1902) rejected the idea of spontaneous generation. What were the alternatives?

Nicolas Steno (Niels Stensen) (1638–1686) argued that the "testes of women is the same organ as the ovary of *Oviparai*. Regnier de Graaf (1641–1673) described testicular tubules as efferent ducts. He recognized ovarian (Graafian) follicles and surmised the function of the Fallopian tubes. He described ectopic pregnancy and assumed that fertilization normally occurs in the ovary. Enrico Sertoli (1842–1910) described the Sertoli cells that line the *tubuli seminiferi contorti* of the testis and nourish developing sperm in 1865.

Jan Swammerdam (1637–1680) realized that egg, larva, pupa, and adult are all different forms of the same animal, *Bybel der natuur* (1668). Lazzaro Spallanzani (1729–1799) was professor at the University of Modena (1760), and Pavia (1768). He successfully fertilized frog eggs *in vitro* and studied the regeneration of the limbs of newts following amputation.

Karl Ernst von Baer (1792–1876) was professor of comparative zoology, then anatomy and physiology at the University of Königsberg (1834–1862). He discovered the mammalian ovum, described the germ layers (ectoderm, mesoderm, and endoderm) as well as the blastula stage and notochord. He maintained that all mammals develop from eggs, *Über Entwicklungsgeschichte der*

Thiere (*About Development History of Animals,* 1828). Baer's law states that "...general characters of the group to which an embryo belongs appear in development earlier than the special characters." His recapitulation theory states that the embryo of a higher animal never resembles the adult of another animal, only its embryo. Although developmental biologists today do not "use" Baer's law, they acknowledge its general validity. Ernst Heinrich Philipp August Haeckel (1834–1919) supported recapitulation — "ontogeny recapitulates phylogeny." His drawings (1874) of embryos of several species might have been embellished to support this assertion.

Matthias Jakob Schleiden (1804–1881) recognized that division of the cell nucleus precedes cell division and is somehow prerequisite to it. Friedrich Leopold August Weismann (1834–1914) described equatorial and reductional divisions in sea urchin eggs. He demonstrated that in multicellular organisms, all inheritance is via germ cells, or gametes; information cannot pass from soma to germ plasm, the "Weismann barrier," *Über die Vererbung* (*On Inheritance,* 1883). Some results from the contemporary studies of horizontal gene transfer and epigenetics indicate slight refinement of this generalization. In 1876, Oskar Hertwig and Herman Fol (independently) injected sperm into sea urchin eggs and observed fusion of their nuclei.

Wilhelm Roux (1850–1924) worked at Jena with Haeckel, at Berlin with Virchow; he was also director of the Institute of Embryology in Breslaus (1879), University of Innsbruck (1889), Breslau (Wroclaw), and director of the Anatomical Institute, Halle (1896–1921). He wrote extensively on development: *Der Kampf der Teile im Organismus* (1881), *Über die Entwicklungsmechanik der Organismen* (1890), *Geschichtliche Abhandlung über Entwicklungsmechanik* (1895), *Die Entwicklungsmechanik* (1905), and *Terminologie der Entwicklungsmechanik* (1912). His own research addressed the development of blood vessels. He ablated, with a hot needle, one or two cells from two- and four-cell stage frog embryos and observed the viability of "half embryos." He explored concepts of natural selection to explain division and growth of some cells and the death, "programmed" in contemporary terminology, of others in normal development. He dissected the medullary plate from the brain of a chicken and maintained it in saline solution. This was the first tissue culture (1885).

Hans Adolf Eduard Driesch (1867–1941) received his Ph.D. from the University of Munich in physics then worked at the Marine Biological Station in Naples in 1891 and published *Analytische Theorie der Organischen Entwicklung* in 1894. He transplanted and reshuffled cells of blastomeres to (re)generate a viable embryo. He introduced the terms "totipotent" and "pluripotent," anticipating current investigations of stem cells. As chair of natural theology at Aberdeen, Scotland, he explored aspects of vitalism in *The Science and Philosophy of the Organism* (1906, see Chapter C8).

Hans Spemann (1869–1941) extended Driesch's experiments, first describing the cell lineage of the parasitic worm, *Strongylus paradoxus*. He removed the lens from a newt's eye and observed its regeneration. He removed cells from embryos by agitation or using a fine noose made from hair, avoiding the artifacts from a hot needle, and transferred nuclei between somatic cells. He described in detail the "fields" of influence of specific cells, including the "Spemann organizer," on the process of development. He was appointed professor of zoology at Freiburg (1919–1937), and was awarded the Nobel Prize in 1935.

Roux, Driesch, and Spemann took physical dissection, injection, and transplantation to the limits of human dexterity. In a sense, the field had to wait for the techniques and concepts of molecular biology and genetics of the 1980s.

C6

Microbiology

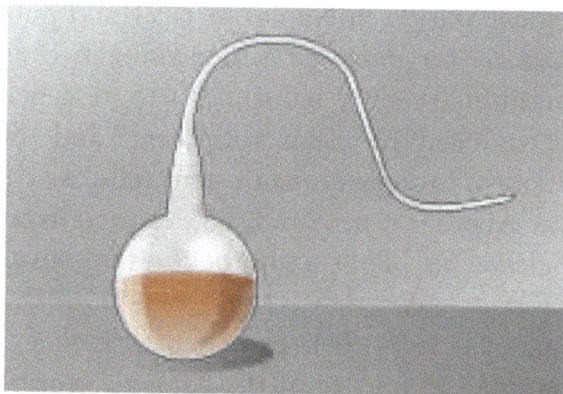

Bouteille en col de cygne ("bottle with a swan neck" after Pasteur).

Pasteur in 1864 compared growth in two boiled broths — one exposed to air and the other not — and concluded: "Never will the doctrine of spontaneous generation recover from the mortal blow struck by this simple experiment."

Diseases of many sorts have been characterized and named in all societies. Their origins have been assigned to various imbalances, transgressions, and spirits. A clear distinction among what we would now call infectious diseases and others — genetic, immune, environmental, psychic, cancer — was made only gradually during the 1800s. Recent advances in genomics indicate that these distinctions are not so clear as was thought only a few decades ago.

The introduction by Jenner of matter from cowpox pustules into the arm of James Phipps hardly meets today's standards of informed consent, but few would condemn the good doctor for exploring variolation. The cell theory, as enunciated by Schleiden, Schwann, and Virchow, was extended to include infectious agents — unicellular micro-organisms, what we would now call eubacteria, archae, and protista.

The germ theory of disease, as enunciated by Koch and others, brought together several concepts. Virchow's dictum, "Omnis cellula e cellula," was essential. The concept first had to be extended to include spores that survived boiling (Pasteurization). The next extension included infectious agents, viruses, that pass through $0.1 \mu m$ filters. As if this weren't bad enough, we now have to contend with *Pandoraviruses*, with 2.5 mega-bases of DNA encoding 1,500 genes, with bits of DNA and RNA that pass from cell to cell via "horizontal gene transfer," and worse yet, with infectious bits of protein, prions. The body summons immune responses on exposure to these "germs" as well as to various protective inoculations. The study of microbiology and immunology remain closely linked to this day.

It is inevitable, and appropriate, that a history of microbiology detail the impacts of infectious diseases. Yet, we now appreciate that on balance we have benefitted immensely from our thousands of co-denizens; even if we do flush a pound of them every day. They long pre-date us; bacteria and protista have been identified in amber over a billion years old.

Thucydides (460–395 B.C.), in the *History of the Peloponnesian War*, described the plague, probably typhus, that killed one-third of the population of Athens (430–410 B.C.) In 1347, Kipchak Khan catapulted bodies of those who had died from the bubonic plague over the ramparts of Kaffa, on the Black Sea, to infect its defenders. Merchants then brought *Yersinia pestis* to Genoa, and the Black Death spread throughout Europe. The plague of 1347–1349 was the worst to strike Europe; it was especially devastating in the north, which carried no residual immunity. In response to the plague of 1665, in which 70,000 died, Londoners killed cats and dogs, only encouraging the proliferation of flea-carrying rats as described by Hans Zinsser in *Rats, Lice, and History* (1935). "Do no harm."

Antonie van Leeuwenhoek (1632–1723) ground 500 lenses, finally making a microscope with 275x magnification. He described protists (*Giardia*), muscle fibers, spermatozoa, blood flow in capillaries, and bacteria in "infusoria."

Smallpox was first recorded in Egypt about 10,000 years ago. The Hittites suffered a small pox epidemic in 1350 B.C. from captured Egyptian soldiers. The Plague of Antonine swept through the Roman Empire during the reign of Marcus Aurelius Antonius. Rhazes in 910 documented immunization against smallpox; it was described in China ~1017. Marseille suffered "Le Grand St. Antoine" in 1720; 80,000 died. Cortez brought smallpox to Veracruz, Mexico, in 1521. Small pox for syphilis; now, that's escalation! In 1763, Sir Jeffrey Amherst gave "smallpox blankets" to the Indians. It is estimated that the population of Native Americans decreased 50 to 80% from 1500–1800 due to diseases, primarily smallpox, introduced by the Europeans. In 1789, smallpox swept through the aborigines of Australia. Japanese aircraft dropped plague-infected fleas in southwest China in 1940. Biological and chemical warfare are now banned to reduce the impact on non-combatants; trinitrotoluene is more humane.

Mary Wortley Montegu, wife of the ambassador to Turkey, introduced smallpox inoculation, as practiced in Turkey, to England on her return in 1721. Zabdiel Boylston inoculated over 200 people, including his 13-year-old son, in Boston, also in 1721. His experiment generated violent opposition from the medical community. He escaped to London, where he published his results as a *Historical Account of the Small-pox Inoculated in New England* (1724).

Jan Ingenhousz (1730–1799) inoculated 700 villagers in Hertfordshire in 1767. Maria Theresa invited him to Vienna in 1768 to inoculate members of her royal family. This success after the smallpox epidemic of 1767 was responsible for changing Austrian physicians' negative view of inoculation and earned Ingenhousz a position as her personal physician.

Edward Jenner (1749–1823) noted that milkmaids who had contracted cowpox did not get smallpox. Following the example of a farmer, Benjamin Jesty, Jenner vaccinated a local boy, James Phipps, with cowpox as described in *An Inquiry into the Causes and Effects of the Variolæ Vaccinæ* (1798), *Further Observations on the VariolæVaccinæ* (1799), *A Continuation of Facts and Observations Relative to the Variolæ Vaccinæ* (1800), and *The Origin of the Vaccine Inoculation* (1801) (vacca, cow). In 1800, Waterhouse gave cowpox vaccination to his son. In 1806, Napoleon ordered that all citizens of France be vaccinated. In 1980 smallpox was declared eradicated by the World Health Organization; one hopes the first of many. Guinea worms are next.

Following, some common infectious diseases and their infective agents:

<u>eubacteria</u>

anthrax	*Bacillus anthracis*, spores
cholera	*Vibrio cholera*
diptheria	*Corynebacterium diphtheriae* (Gram +)

gonorrhea	*Neisseria gonorrhoeae*
leprosy = Hansen's disease	*Mycobacterium lepra*
plague	*Yersinia pestis*
syphilis	*Treponema pallidum*
tuberculosis	*Mycobacterium tuberculosis,*
	M. bovis, M. africanum
typhoid fever	*Salmonella enterica* (Gram −)
typhus	*Rickettsia prowazekii* (Gram −)
whooping cough	*Bordetella pertussis*

<u>protista</u>

giardia	*Giardia lamblia*
malaria	*Plasmodium falciparum*
	(Anopheles gambiae)
scabies	*Sarcoptes scabiei* (mite)

<u>viruses</u>

Aids	Human Immunodeficiency Virus
flu	Influenza A virus "Spanish,"
	subtype H1N1
hepatitis A	Hepatitis A virus
herpes	Herpes simplex virus
influenza	Orthomyxoviridae
poliomyelitis	Polio virus
rabies	Lyssa virus
small pox	Variola (Homo) Vaccinia (Bos)
	virus
yellow fever	Flaviviridae, +ssRNA arbovirus;
	(Aedes simpsaloni)

Jean-Baptiste Lamarck (1744–1829) considered that spontaneous generation was ongoing and accounted for seemingly unexplainable fermentations and contaminations. Theodor Schwann (1810–1881) referred to "factors in the air" that caused fermentation in 1837.

John Snow (1813–1858) established epidemiology with his *On the Mode of Communication of Cholera* (1849):

> On proceeding to the spot, I found that nearly all the deaths had taken place within a short distance of the pump. There were only ten deaths in houses situated decidedly nearer to another street-pump. In five of these cases the families of the deceased persons informed me that they always went to the pump in Broad Street (London), as they preferred the water to that of the pumps which were nearer. In three other cases, the deceased were children who went to school near the pump in Broad Street…

As elaborated in *Medical Times and Gazette*:

> It was discovered later that this public well had been dug only three feet from an old cesspit that had begun to leak fecal bacteria. A baby who had contracted cholera from another source had its diapers washed into this cesspit, the opening of which was under a nearby house that had been rebuilt farther away after a fire had destroyed the previous structure, and the street was widened by the city. It was common at the time to have a cesspit under most homes. Most families tried to have their raw sewage collected and dumped in the Thames to prevent their cesspit from filling faster than the sewage could decompose into the soil.

Ignaz Phillip Semmelweis (1818–1865) practiced in the Vienna Allgemeines Krankenhaus (General Hospital). He proposed that "particles" cause puerperal fever (L. *puer* "child, boy") and was spread by doctors. He was appointed chair of midwifery in St. Rochus Hospital in Pest, Hungary (1855). He introduced a chlorine solution to wash instruments, bedding, and clothing; the incidence of puerperal fever fell from 18.3 to 1.3%, *Die Ätiologie, der Begriff und die Prophylaxis des Kindbettfiebers* (1861). His practice was not accepted in Austria.

Rudolf Ludwig Karl Virchow (1821–1902) had championed *Omnis cellula e cellula* and *Omne vivum ex ovo*, and rejected spontaneous generation. He wrote in his *Report on the Typhus Outbreak of Upper Silesia* (1848): "… The outbreak could not be solved by treating individual patients with drugs or with minor changes in food, housing, or clothing laws, but only through radical action to promote the advancement of an entire population." He opposed the theory that bacteria cause disease and rejected Semmelweis' chlorine wash. Hey, you can't get them all right!

Louis Pasteur (1822–1895) exposed boiled (sterilized) broths to air and demonstrated that the infective agent could be excluded by a fine filter. He also demonstrated, using his "bouteille en col de cynge," that the oxygen required for growth in the broth could diffuse over the length of the "swan neck"; however, microbes could not. He developed techniques for pasteurization of wine and milk; he endorsed antiseptic surgery as developed by Lister. He immunized chickens with a "weakened" cholera and immunized cattle against anthrax (1879). He endorsed the germ theory of disease. He developed an anti-sera against the lyssa virus and tested it in 11 dogs before treating Joseph Meister, who had been bitten by a rabid dog in 1885. Meister was employed as gatekeeper at the Institut Pasteur until his death in 1940.

Joseph Lister (1827–1912) received his Bachelor of Medicine in 1852. Motivated by Pasteur's work, he described his *Antiseptic Principle of the Practice of Surgery* in 1867. It included wearing clean gloves as well as washing hands and instruments with 5% carbolic acid (phenol). He acknowledged: "Without Semmelweis (and Holmes), my achievements would be nothing." One might wish that such integrity and humility were as contagious as the *Strep.* and *Staph.* infections with which he struggled.

Ferdinand Julius Cohn (1828–1898) found that some bacteria survive sterilization by boiling. He characterized the vegetative state and the endospore that is more resistant to hostile environments. This lead to the practice of boiling at elevated temperature at several atmospheres pressure. Gerhard Henrik Armauer Hansen (1841–1912) received his M.D. from the University of Oslo. In Bergen, Norway, he identified in 1873 *Mycobacterium leprae* in tissues of patients suffering from leprosy, now called Hansen's disease.

Robert Koch (1843–1910) worked in the Imperial Health Office, Berlin (1880), and became a professor in Berlin in 1885. He identified *Bacillus anthracis* in 1877, *Mycobacterium tuberculosis* in 1882, and *Vibrio cholera* in 1883. Koch's postulates (1882) are still the gold standard but cannot always be realized when establishing the cause of an infectious disease.

The (micro)organism is:

1) found in all cases of the disease examined.
2) prepared and maintained in a pure culture.
3) capable of producing the original infection, even after several generations in culture.
4) retrievable from an inoculated animal and cultured again.

Some rejected Koch's findings, as others had dismissed Semmelweis. A colleague from München wrote:

> Herr Doctore Pettenkofer presents his compliments to Herr Doctor Professor Koch and thanks him for the flask containing the so called *Cholera vibrios*, which he was kind enough to send. Herr Doctore Pettenkofer has now drunk the entire contents and is happy to be able to inform Herr Doctor Koch that he remains in his usual good health.

Biological systems are tough!

Paul Ehrlich (1854–1915) was a colleague of Koch in Berlin and subsequently worked at the Institut Serumforschung und Serumprüfung (Serum Research and Testing, 1896), and at the University of Göttingen (1904). He developed salvarsan, the first sulfa drug, the original "magic bullet," to treat syphilis and predicted autoimmunity, "horror autotoxicus."

Inevitably discussions of microbiology emphasize diseases and leave the impression that germs are bad. Quite the contrary, ecosystems are absolutely dependent on recycling by soil bacteria as are the beverage and cheese industries. The vast majority of bacteria found in and on man and beast benefit their host. Just as people have fingerprints, so too they have unique micro-biomes. In contrast to the invariance of fingerprints, these microbe signatures can change with circumstances; yet two people in nominally identical environments have slightly different signatures.

C7

Pharmacology

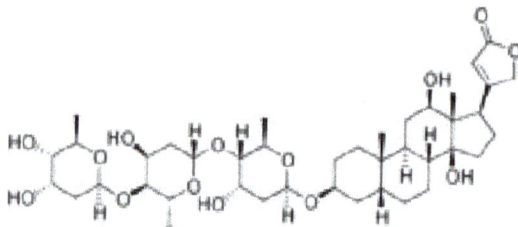

Foxglove (*Digitalis*) Digoxin.

Given the close interactions between medicine and biology, it is hardly surprising that the basic medical sciences of most medical schools are organized around six departments — anatomy & cell biology & embryology, physiology, microbiology & immunology, pharmacology, biochemistry, and neurosciences. Endocrinology sometimes gets its own department. This organization approximately reflects historical development. These divisions become ever more arbitrary as the concepts and techniques of contemporary cell and molecular biology tend to unite all of these fields.

Most cultures have home remedies, many based on plant extracts. This folk wisdom provided quinine, aspirin, digitalis, and taxol, and is still a source of "lead" compounds for drug development. Ever more extensive pharmacopeias were published with the passing centuries; their documentation and illustrations were one of the driving forces in the development of botany.

In parallel, some alchemists explored the therapeutics of non-botanics, usually inorganic compounds, but sometimes elements. An overdose of some herbal might be unpleasant; mercury or sulfur could be lethal. Finally, in the mid-1800s organic chemists, especially in Germany, began to synthesize new compounds, often derived from dyes, and test them in animals as candidate drugs.

Until the past decade, most drugs came from plants or fungi. Today pharmacology is changing rapidly because the search for leads can be done by so-called click chemistry, in which large matrices of reactants can be mixed and monitored then tested in standard tissue cultures or cell suspensions; all done robotically.

The distinction between pharmacology and biochemistry (Chapter C8) is, in concept, clear. The former treats the effects on body and mind of compounds whose origin is external to the body. Biochemistry explores the chemical processes that occur within the body, eukaryote or prokaryote. In practice and in history this division is not so clear. All of these reductionist divisions (Chapters C3–C8) tend to merge as one explores cellular and molecular mechanisms.

Pedanius Dioscorides ($\sim$40–$\sim$90 B.C.) cataloged 500 plants and $\sim$1600 "drugs" in the five volumes of his *De materia medica*. This set the pattern for well over a millennium. Within the prevailing conceptual framework, Galen (129–$\sim$210) was correct in ascribing illness to an imbalance of the four humors.

Paracelsus (1493–1541) voiced the interests of many alchemists (Chapter B5): "Many have said of Alchemy, that it is for the making of gold and silver. For me such is not the aim, but to consider only what virtue and power may lie in medicines." As chair of medicine at the University of Basel (1536), he succinctly summarized the essence of "dose, response" — *"Alle Ding' sind Gift und nichts ohn' Gift; allein die Dosis macht, dass ein Ding kein Gift ist."* "All things are poison and nothing is without poison; only the dose permits something not to be poisonous."

Nicholas Culpeper (1616–1654) operated a pharmacy at Spitalfields near London where he examined poor patients and dispensed herbal supplies for free. He examined patients in person and famously noted ". . . as much piss as the Thames might hold . . ." would not aid in diagnosis. He regarded medicine as a public asset; not a commercial secret. "Three kinds of people disease the people — priests, physicians and lawyers." He dispensed foxglove, which contains digoxin, to treat heart conditions (to increase stroke volume, decrease pulse, and increase $[Ca^{2+}]_{cytosol}$). "Culpeper" herb and spice shops, were found throughout Britain. His *Complete*

Herbal (1652), *The English Physician* (1653), and *Philosophy of Herbalism; Alternative Medicine* went through many editions.

Erasmus Darwin, grandfather of Charles, published *An Account of the Successful Use of Foxglove in Some Dropsies and in Pulmonary Consumption* in 1785. "Whilst the last pages of this volume were in the press, Dr. Withering of Birmingham . . . published a numerous collection of cases in which foxglove has been given, and frequently with good success." William Withering (1741–1799) was a physician at Birmingham General Hospital. He described the treatment of patients with (hy)dropsy with extracts of foxglove and subsequent increase in urine flow and reduction of edema.

Friedrich Sertürner characterized morphine as the first active alkaloid extracted from the opium poppy plant in 1804. As illustrated by these examples of alchemy and foxglove, most early, successful pharmaceuticals came from plants; they are still the source of lead compounds for drug development (Chapter C10). Many developing countries in Africa, Southeast Asia, and Latin America have become much more protective about "explorers" collecting plants that might yield drugs.

Humphrey Davy (1778–1829) described N_2O, nitrous oxide, as laughing gas. However, he was quite busy with other chemical pursuits (Chapters B6, B7, and C8) and did not explore its clinical use as an anesthetic. Ascanio Sobrero (1812–1888) was appointed professor of chemistry at the University of Turin in 1845. His discovery of (tri)nitroglycerin lead to insights in vasodilation — what better example of the unintended consequences of research? Alfred Bernhard Nobel (1833–1896) mixed the dangerously explosive nitroglycerin with diatomaceous earth (kieselguhr) to make dynamite, patented in 1867. It transformed the mining and related industries and made a fortune for Nobel. The basic reaction is:

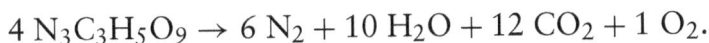

$$4\ N_3C_3H_5O_9 \rightarrow 6\ N_2 + 10\ H_2O + 12\ CO_2 + 1\ O_2.$$

This is an explosion — immediate release of over seven equivalents of hot gas — in contrast to the rapid oxidation of gun powder. The gun powder and related explosives, as trinitrotoluene (TNT), were also put to military use. In response to an erroneous death notice, a Paris newspaper published the headline "Le marchand de la mort est mort." (The merchant of death is dead.) This perception provided some of the motivation for his creation of the Nobel Peace (and other) Prizes.

Nobel suffered from heart pains, angina pectoris, probably from restricted blood flow in the coronary artery. "My heart trouble will keep me here in Paris for another few days at least, until my doctors are in complete agreement about my immediate treatment. Isn't it the irony of fate that I have been prescribed N/G 1, (nitroglycerin) to be taken internally! They call it Trinitrin, so as not to scare the chemist and the public." In simplification, mitochondrial aldehyde dehydrogenase releases nitric oxide (NO) from nitroglycerin; this increases the concentration of cGMP (cyclic guanosine-mono-phosphate in smooth muscle cells lining blood vessels; this, in turn, increases calcium entry, which relaxes smooth muscles with subsequent increase in blood flow, relieving the angina (after Ferid Murad, Nobel Laureate, 1998). Viagra functions by a related mechanism; early, non-selective vaso-dialators produced a (sometimes) unwelcome side effect.

Paul Ehrlich (1854–1915) worked at the Institute for Infectious Diseases in Berlin (1891), with Koch (Chapter C6), at the Institut Serumforschung und Serumprüfung (Institute for Serum Research and Serum Testing, 1896), and the University of Göttingen (1904). He developed salvarsan (arsphenamine) and a family of sulfa drugs to treat syphilis and coined the phrase, "magic bullet." He also predicted "horror autotoxicus," or autoimmunity; that is, a pathologic immune response to the body's own tissues.

Gerhard Domagk (1895–1964) was director of Bayer's Institute of Pathology and Bacteriology in Wuppertal in the 1930s where he developed sulfonamidochrysoidine (prontosil), the first commercially available antibiotic. He was awarded the Nobel Prize in 1939, but was allowed by the German government to accept it only in 1947.

Alexander Fleming (1881–1955) described the enzyme lysozyme as the "body's own antibiotic" in 1922 and blurred the characterization of pharmacology as the study of drugs from outside the body. He was the first to record in 1929, though perhaps not the first to observe, a zone of bacterial clearing on an agar plate caused by *Penicillium notatum*. Howard Walter Florey (1898–1968) received his M.D. from the University of Adelaide in 1921, and his Ph.D. from Cambridge in 1927. He was appointed professor of pathology at Oxford in 1935, where he and Ernst Boris Chain worked on the extraction and purification of penicillin. Its use aided the Allies' war efforts. It replaced sulfonamides for most general applications. Fleming, Florey, and Chain shared the Nobel Prize in 1945. Fleming returned home to assume the chancellorship of the Australian National University (1965–1968). He was disarmingly frank: "People sometimes think that I and the others worked on penicillin because we were interested in suffering humanity. I don't think it ever crossed our minds about suffering humanity. This was an interesting scientific exercise, and because it was of some use in medicine is very gratifying, but this was not the reason that we started working on it."

The search for new drugs, "pharmaceuticals," is big business today. How to evaluate its success and who should pay its costs and reap its benefits becomes an evermore important concern.

C8

Biochemistry

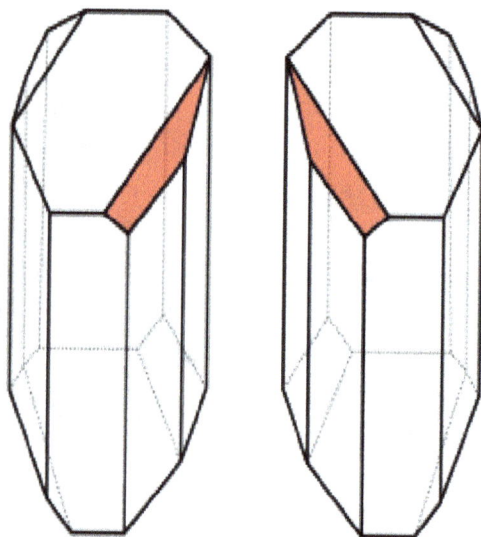

Tartaric acid-crystal of D- and of L-enantiomorphs
of tartaric acid (after Pasteur).

Just as organic chemistry developed as a distinct branch of chemistry, so did biochemistry develop from physiology and pharmacology. Just as physicists in the late 1700s wondered whether Galvani's animal electricity was different from Volta's "battery," so biologists had to ponder the vital nature of biological molecules. The inadvertent synthesis of urea $(CO(NH_2)_2)$ from ammonium cyanate $(NH_4^+ OCN^-)$ by Wöhler in 1828 sounded the death knell for *vis vitalis* and vitalism in general. However, the patient was quite resilient and lingered on for over a century. Organic chemists applied their concepts and techniques to study the molecules of life.

Pharmacologists initially focused on characterizing compounds of external origin that elicit a specific physiological or behavioral response. Biochemists addressed a broader spectrum of "inherent" chemical reactions in eukaryotes and bacteria. Sorting out the pathways of "intermediary metabolism" was one of the great achievements of biochemistry. A seemingly closed chapter has been reopened at higher resolution under the rubric of metabolomics. The distinction between cell biology, pharmacology, and biochemistry becomes ever less meaningful as the concepts and techniques of molecular biologists sort out pathways and mechanisms.

As is so often the case when considering the history of science, the story is best told through the achievements of distinguished individuals. They worked in a stimulating intellectual environment, fully as important, but more difficult to characterize.

Jan Baptist van Helmont (1577–1644) described "gas sylvestre" (CO_2) from burning charcoal and suggested the analogy with fermentation. He grew a willow in a pot of soil, with the same weight before and after, to 75 kg and wondered how that mass might have come from water. He proposed that digestion is aided by a "ferment" in the stomach and recommended alkalins for acid stomach. He posed questions for the scientific revolution that would be answered only in the 20th century.

Jan Ingenhousz (1730–1799) observed that submersed plants in light give off bubbles, later identified as oxygen (Chapters B6, C10); in shade, the bubbles stop. In the dark, plants give off carbon dioxide, *Experiments upon Vegetables Discovering their Great Power of Purifying the Common Air in the Sunshine: An Essay on the Food of Plants and the Renovation of Soils* (1796). Joseph Priestly (1733–1804) characterized several "airs" (Chapter B6) and their effects on mice. Antoine Lavoisier concluded that organic compounds must be composed of carbon, hydrogen, and oxygen (~1780).

William Beaumont (1785–1853) served as a surgeon's mate in the U.S. Army (1812–1815 and 1819–1821). Alexis St. Martin, working for the American Fur Company on Mackinac Island, Michigan, was accidentally shot in the stomach in 1822. The surrounding tissue healed; however, the stomach tissue fused with the body wall and the fistula remained open. For many years Alexis served as a handyman for Dr. Beaumont, who could monitor the digestive process by inserting food on a string and extracting digestive juices from Alexis's stomach. One is reminded of Galen's comment on the wounds of gladiators providing "windows into the body" (Chapter C2).

Friedrich Wöhler (1800–1882) was a "post-doc" with Jöns Jakob Berzelius (Chapter B7) in Stockholm. Almost by accident, he synthesized the organic compound, urea ($CO(NH_2)_2$), from inorganic, ammonium cyanate ($NH_4^+ OCN^-$) in 1828. Although this falsifying event should have spelled the end of *vis vitalis* and vitalism (Chapter A13), these concepts survived for over a century. Wöhler was appointed professor of chemistry at the University of Göttingen in 1836.

Claude Bernard (1813–1878) worked as a preparateur at Collège de France (1841), and was appointed professor (1855), then professor (1868) at the Muséum national d'Histoire naturelle. He investigated the role of pancreatic juice in digestion, the uptake of sugar from the blood by the liver, *Lessons on Diabetes and Animal Glycogenesis* (1877), the effects of curare and of carbon monoxide, and the functions of vaso-dilator and vaso-constrictor nerves. He was both insightful and succinct: "The account of the liver shows very clearly that there are *internal secretions*, that is to say, secretions whose products, instead of being poured out to the exterior, are transmitted directly into the blood." "Experiment is fundamentally only induced observation." "The true scientist is one whose work includes both experimental theory and experimental practice ... *a priori* conception ... & ... *a posteriori* interpretation ..." He noted, long before Popper (Chapter A10), that "... we must never make experiments to confirm our ideas, but simply to control them." "... we must no longer defer to authority, as scholasticism did." He took a broad view of the philosophy of science: "... the application of mathematics to natural phenomena is the aim of all science ..."; however, he cautioned against inappropriate use. His *Introduction to the Study of Experimental Medicine* (1865) was still a revered text in the early 1900s.

The work of Louis Pasteur (1822–1895) spans and combines many fields (Chapters C4, C6, and C7). He noted that tartaric acid of biological origin rotates polarized light, whereas that of chemical origin does not (1849). Crystals of "chemical" origin have a visible handedness. Those of one hand rotate light as does tartaric acid of biological origin; those of the other hand rotate in the opposite direction. His discovery of chirality was rationalized with the discovery of tetrahedral (sp^3 orbitals) carbon coordination. Joseph Lister (1827–1912) acknowledged that his use of 5% carbolic acid (phenol) to wash hands and instruments was motivated by Pasteur, perhaps more an acknowledgement of intellectual environment than of a specific experiment.

Frederick Grant Banting (1891–1941), in collaboration with a graduate student, Charles Best, extracted insulin from the pancreas of dogs and discovered that it caused release of glucose, stored in the liver. The division between physiology, pharmacology, endocrinology, and biochemistry was thus further blurred. He and John James Rickard Macleod were awarded the Nobel Prize in Medicine in 1923.

From hormones to pheromones, biochemistry knows no bounds. Male goats secrete 4-ethyloctanal, which converts to 4-ethyloctanoic acid, responsible for that "goaty odor." It both attracts females and starts a hormonal chain reaction in the ewe's brain that triggers ovulation. It will soon be available in after-shave lotion.

C9

Neurobiology

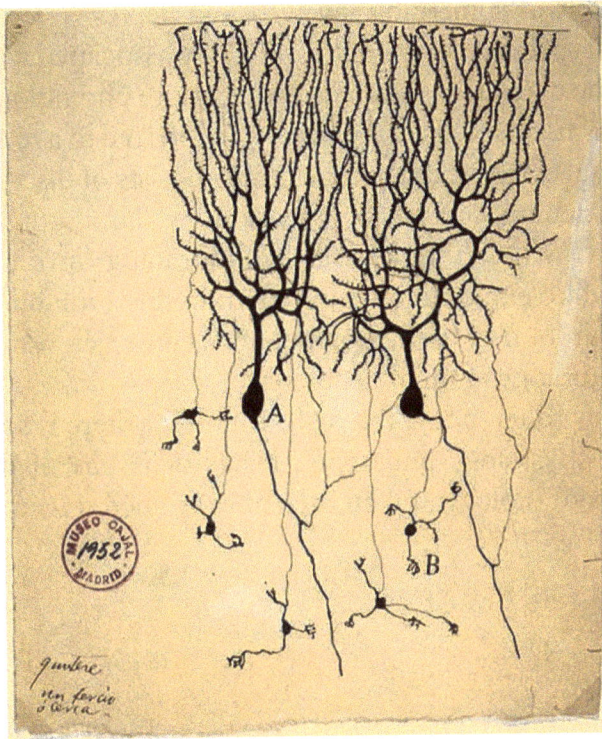

Purkinje cells
(after Cajal, 1899).

Some contended that the heart is the source of thought and emotions; this reference to "heart" is deeply imbedded in our literary traditions. Others favored the brain. Galen referred to the lost works of Herophilos, who argued that the brain is the center of nervous system and site of intelligence. He distinguished nerves from blood vessels, no trivial achievement before the days when cadavers were injected with latex to facilitate dissections, and he supposedly distinguished motor from sensory nerves. Ibn al-Nafis identified 10 cranial nerves and noted that "...each nerve (of the eye) goes to the opposite side."

Many hunters and farmers had observed the twitching of recently killed animals. Galvani was the first to systematically explore the stimulus of the nerves of dead frogs and the resultant contraction of their muscles. These initial stimuli were mechanical, then electrical; there was something special about nerves.

Gall showed that specific regions of the brain are associated with specific functions; he extended his research to explore phrenology. Golgi had identified distinct cells in the brain and referred to a reticulum. Ramón y Cajal established that the brain consists of discrete cells and that the reticulum consists of extensions, axons, that contact other cells. Pavlov described classical conditioning (involuntary reflex); Thorndike explored operant conditioning (voluntary reflex). The triumvirate of Frisch, Lorenz, and Tinbergen established the field of ethology, animal behavior.

Although obvious to us today, the realization that there is specific anatomy and physiology underlying behavior is one of the major achievements of biology and an area of continued research.

Galen (129–~210) referred to the lost works of Herophilos (335–280 B.C.), who argued that the brain is the center of the nervous system and site of intelligence. Galen distinguished nerves from blood vessels, no trivial achievements before the days when cadavers were injected with latex to facilitate dissections, and he supposedly distinguished motor from sensory nerves. Ibn al-Nafis (1213–1288) identified 10 cranial nerves and noted that "...each nerve (of the eye) goes to the opposite side." He maintained that cognition, sensation, imagination, and locomotion come from the brain.

Jan Swammerdam (1637–1680) "irritated" the (motor) nerve of a dead frog and observed its leg twitch. He removed its heart and touched areas of the brain to the same effect. In *Bybel der natuure* (1668), he wrote: "...motion or irritation of the nerve alone is necessary to produce muscular motion"; no "mystic spirit" need be invoked. Luigi Galvani (1737–1798), using electricity from a Leyden jar, sparked the muscles of dead frogs and observed them twitch in 1771. Thus began the discussion of biological vs. non-biological electricity. Alessandro Giuseppe Antonio Anastasio Volta (1745–1827) was appointed professor of experimental physics at the University of Pavia in 1779, then professor of philosophy at Padua in 1815. His invention of the voltaic pile battery — alternating layers of sheets of metal, e.g. zinc and copper, separated by paper or cloth saturated with an electrolyte — in 1799 was copied and refined in many labs and provided a reliable source of direct current for experiments and application (*Volta: Science and Culture in the Age of Enlightenment*, 2003).

Thomas Young (1773–1829) was a true polymath as inferred in the title of the recent biography, *The Last Man to Know Everything* (2006). Maxwell praised him: "Thomas Young was the first who, starting from the well-known fact that there are three primary colors, sought for the explanation of this fact, not in the nature of light but in the constitution of man." We now describe color vision in terms of cone cells and their three photopsins. In his first paper, read to the Royal Society at age 20, he presented his analysis of visual accommodation:

> It is well known that the eye, when not acted upon by any exertion of the mind, conveys a distinct impression of those objects only which are situated at a certain distance from itself; that this distance is different in different persons, and that the eye can, by volition of the mind, be accommodated to view other objects at a much lesser distance; but how this accommodation is effected, has long been a matter of dispute, and has not been satisfactorily explained.

Hermann Ludwig Ferdinand von Helmholtz (1821–1894) invented the ophthalmoscope, an immediate success, *Handbuch der physiologischen optik* (1851). He measured the velocity of a nerve impulse along a frog axon (~100 m/sec).

Ivan Petrovich Pavlov (1849–1936) first described "psychic secretion"; rats and dogs salivate prior to being fed. If a bell is always rung prior to feeding, the dogs salivate at the sound of the bell in the "conditional or classic reflex." He also described the involuntary reactions to stress and pain and identified the nerves that control gastric secretions for which he was awarded the Nobel Prize in 1904.

Santiago Ramón y Cajal (1852–1934) was appointed director of the Zaragoza Museum in 1879 and director of the National Institute of Hygiene in 1899. Using a silver chromate stain, first used to view brain tissue by Camillo Golgi (1843–1926), he described the nervous system as consisting of billions of separate neurons, not a reticulum as interpreted by Golgi. Ramón y Cajal and Golgi shared the Nobel Prize in 1906.

Sigmund Freud (1856–1939) named and characterized the unconscious, the Oedipus complex, defense mechanisms, Freudian slips, dream symbolism, etc. Although many of his techniques and conclusions are now out of favor, he had an enormous impact on the development of psychoanalysis.

Charles Scott Sherrington (1857–1952) received a Bachelor of Medicine and Surgery from Cambridge in 1885 and then studied with Koch and with Virchow. He identified the muscle spindles fibers that initiate the stretch reflex and described reciprocal innervation, that is, groups of muscles that have an excitation — inhibition relationship. He described the postural reflex, the anti-gravity stretch reflex, and proprioception. He shared the Nobel Prize with Edgar Adrian in 1932.

Edward Lee Thorndike (1874–1949) was elected president of the American Psychological Association in 1912 and co-established the journal *Psychometrika* in 1937. He and Skinner used punishments and rewards to teach cats to traverse mazes. As described in *The Elements of Psychology* (1905) and *Animal Intelligence* (1911), he introduced the term "operant conditioning," to contrast with the "conditional learning" of Pavlov.

The Nobel Prize in 1973 was awarded to Karl Frisch, Konrad Lorenz, and Nikolaas Tinbergen; they had laid the foundations of ethology. Karl Ritter von Frisch (1886–1982) was professor of zoology at the Universities of Rostock, Breslau, Munich, and Graz. He studied the behavior of the bee, *Apis mellifera carnica*. He described pheromones from the queen bee, the bees' sense of direction by polarization of light and elevation of the sun, orientation of the honeycomb, as well as the round dance and the waggle dance that communicate to fellow bees the location of a food source. Konrad Zacharias Lorenz (1903–1989) was professor of psychology when he was drafted into the Wehrmacht in 1941. He served as a medic, was captured and held as a prisoner of war in the U.S.S.R. from 1942–1948. After returning to Germany he worked at the Max Planck Institute for Behavioral Physiology in Seewiesen. He studied fixed patterns of action or instinctive behaviors and described imprinting in young animals. He "... realized that an overpowering increase in the drives of feeding as well as of copulation and a waning of more differentiated social instincts is characteristic of very many domestic animals." He feared "... that analogous processes of deterioration may be at work with civilized humanity." Such interpretations lent support to racial preferences (Chapter C15). Nikolaas Tinbergen (1907–1988) worked at the University of Leiden and was a P.O.W. in Germany during the war. He asked four questions of behavior that still offer guidelines today:

1) What are the stimuli that elicit the behavior?
2) How does the behavior change with age?
3) How does the behavior compare with similar behaviors in related species?
4) How does the behavior impact survival and reproduction?

Alan Lloyd Hodgkin (1914–1998) and Andrew Fielding Huxley (1917–2012) performed many voltage clamp experiments on the giant axon of the giant squid, *Loligo pealei*. They hypothesized ion channels before they were characterized by biochemists and the "Hodgkin, Huxley model of axonal transmission" in 1952, for which they were awarded the Nobel Prize in 1963. Their model involves the parameters: C_m capacitance of the lipid bilayer, I_i (transmembrane) ionic currents, V_m transmembrane potential, E_i reversal potential, g_i channel conductance, g_n maximum conductance, $\phi_\alpha \chi_\beta$ gating variables, $\tau_\phi \tau_\chi$ time constants for activation and inactivation. Most of these empirical constants can now be related to molecular properties of the channel proteins.

By 1859 enough was known of animal physiology in general and related disciplines outlined in Chapters C3–C9 so that Darwin could see the patterns that should be encompassed by a general theory. Subsequent advances were best understood in the context of evolution by natural selection.

C10

Botany

Chriftwurtz.

Helloborus niger from *Herbarum vivae eicones* (Otto Brunfels).

Much of the early interest in botany was motivated by the use of plants as the sources of drugs — digitalis, opium, cannabis, aspirin, etc. Witness the succession of pharmacopeias.

Those students of natural history who truly loved plants and were fascinated by their endless variety and beauty prided themselves on identifying the various species and their subtle variations. Even today, some botanists resent having to justify their research in terms of drug discovery or to compare their classifications with those based on DNA sequences. Bully for them!

Given the successes of a reductionist approach to science in general and biology in particular, it requires a fundamental change of perspective to appreciate the value of "proper" characterization and classification. Lord Kelvin infamously quipped: "Physics is science; the rest is stamp collecting." He just didn't get it.

This initial survey employs two over-simplifications. Botanists were most concerned with proper classification. Not true; early on several men addressed mechanistic questions of plant physiology. All other biologists focused on understanding mechanisms. Not true; zoologists and paleontologists were equally concerned with a "natural" system of classification, to be elaborated in Chapter C13.

Getting the right ordering, trivial as it may sound, was fundamental to the advancement of astronomy and of chemistry. That right ordering is what underlies Mendeleev's periodic table and subsequently Bohr's model of the atom. The concepts upon which that ordering is based are fundamentally different from those of the binary classification of organisms by Linnaeus.

Helmholtz and Warming extended the study of botany to biogeography and ecology.

Theophrastus (370–285 B.C.) was a student of Leucippus and became the successor of Aristotle as leader of the peripatetic school in Athens. His *Enquiry into Plants* (nine books) and *On the Causes of Plants* (six books) listed most of the plants known to the Greeks. He is considered to be the father of taxonomy. He was opposed to Aristotle's "Great Chain of Being" (Chapter C15). Padanius Dioscorides (~40–~90) compiled the first pharmacopeia, listing 500 plants, *De materia medica* (five books).

Otto Brunfels (1488–1534) was a physician in Bern. His *Herbarum vivae icons* (three books, 1530–1536) and *Contrafayt kräuterbuch* (two books, 1530 and 1536) were, at the time, the most complete listings of European plants; the names were given in the vernacular, German. Hieronymus Bock (1498–1554) studied at the University of Heidelberg. He became caretaker of the grounds of the Count Palatinate of Zweibrücken for nine years, then prince's physician in 1532. His *Kreuterbuch* (*Plant Book*) described 700 species. Leonhart Fuchs (1501–1566) became chair of medicine at Ingolstadt in 1526 and personal physician to Georg, Margrave of Brandenburg, from 1528–1531. He established the first botanical garden in Germany.

Konrad Gessner (1516–1565) received his M.D. at Montpellier in 1541 and lectured in physics in Zürich. He compiled known life, including specimens from the Americas in *Enchiridion historiae plantarum* (*Handbook of the History of Plants*) in 1541, *Catalogus plantarum* (in four languages) in 1542, as well as *Historiae animalium* (1551–1558), which began modern zoology.

Pierre Belon (1517–1564) is not in the pantheon of botanists; however, he traveled throughout Asia Minor, Egypt, Arabia, and Palestine and brought the perspective of biogeography before the term was invented. His writings include: *Histoire naturelle des estranges poisons* [fish] (1551) *De aquatilibus* (1553), *Les observations de plusieurs singularite et choses memorables trouvées en*

Grèce, Asie, Judée, Egypte, Arabie et autres pays étrangèrs (1553), and *L'Histoire de la nature des oyseaux* [birds] (1555). After all these travels he was murdered in Bois de Boulonge, back home in Paris.

Andrea Cesalpino (1519–1603) established a formal garden in Padua in 1546. He was director of the botanical garden in Pisa (1554–1558). His *De plantis libri* XVI (1583) was the first systematic classification of flowering plants based on their reproductive organs; he recognized 15 "higher genera." In the 1590s, he served as physician to Pope Clement VIII.

John Ray (1627–1705) made several plant and animal collecting tours in the Netherlands, Germany, Italy, and France. He classified plants by mutual fertility: *Catalogus plantarum Angliae* (1670 and 1677), *Methodus planiarurn nova* (1682), *Historia generalis plantarum* (3 Vols, 1686, 1688, 1704), and *Historia Plantarum* (1686–1704) referencing 18,600 plants. He included animals: *Synopsis methodica Animalium Quadrupedum et Serpentini Generis* (1693), and *Synopsis methodica avium et piscium* (1713). Most of these botanists were aware of the works of their colleagues; however, there was no formal organization or curriculum.

Nicholas Culpeper (1616–1654) documented the use of herbals (Chapter C8), *Philosophy of Herbalism; Alternative Medicine and Complete Herbal* (1652). Robert Hooke (1635–1703) coined the term *cellula* to describe the entities that he observed in cork with his 3 × lens (Chapter C4) in *Micrographia* (1665).

Many, if not most, drugs or pharmaceuticals are secondary metabolites of plant origin; many are inferred to protect the plants from parasites. For many, neither the mechanism of action nor the side effects have been fully explored (Chapter C7). A few of the thousands are listed. As noted, many tropical nations strictly regulate the collection and export of plants; they are keen to receive a share of any profits that might be realized from the subsequent

development of a product. Additionally, several drugs, e.g. penicillin, come from fungi.

active ingredient	source	antiquity
aloe vera	*Aloe vera*	
aristolochic acid	birthwort, *Aristolochia*	China
artemisinin	*Artemisia annua*	China
aspirin, salicylic acid	white willow, *Salix alba*	Egypt
astragalus	*Astragalus propinquus*	China
belladonna	*Atropa belladonna*	
capsaicine	chili pepper, *Capsicum frutescens*	
coffee senna	*Cassia occidentalis*	
digitalis, digoxin	foxglove, *Digitalis lanata*	
ephedra	*Ephedra sinica*	China
garlic	*Allium sativum*	
ginseng	*Panax ginseng*	China
inulin	dahlias	
morphine, opium, codeine	poppy, *Papaver somniferum*	
quinine	*Chichona*	
taxol	fungus on *Taxus brevifolia*	

Van Helmont, Hales, and Ingenhousz addressed questions
of plant and animal physiology. Jan Baptist van Helmont
(1577–1644) made a range of experiments, e.g. on digestion (Chapter C9). He demonstrated that the mass of a tree grown in a pot
of soil took its weight from the air and water, not from the soil;
this observation led to the study of photosynthesis. Stephen Hales
(1677–1761) described the loss of water from plants by evapora-
tion in *Vegetable staticks* (1727) and his observations on the circulatory system in *Haemastaticks* (1737). Jan Ingenhousz (1730–
1799) was personal physician to Austrian Empress Maria Theresa;

he inoculated her and her family against small pox (Chapter C6). He also recorded that in light, plants give off bubbles, which he identified as oxygen; in shade no bubbles were emitted; in the dark they give off carbon dioxide. He recorded, as previously noted by van Helmont, that the mass of a plant comes from the air, not the soil, *Experiments upon Vegetables Discovering their Great Power of Purifying the Common Air in the Sunshine.* Ferdinand Julius Cohn (1828–1898), like many of his contemporaries, considered bacteria to be plants (Chapter C6). He described bacterial spores that survived boiling at one atm., 100°.

Carl Linnaeus (1707–1778) was appointed professor of botany at Uppsala in 1741. He introduced a binary nomenclature, genus and species, still used today in a sequence of editions of *Systema naturae*, first edition (1735), to 10th edition (1758), with 7,700 species of plants and 4,400 species of animals. In his binary names for plants the word *officinalis* indicated medicinal use.

Friedrich Wilhelm Heinrich Alexander Freiherr von Humboldt (1769–1859) had a broad range of interests, including geology, *Mineralogische Beobachtungen über einige Basalte am Rhein (Mineralogical Observations of a Unique Basalt on the Rhein,* 1790). He traveled throughout South America, 1799 to 1804, making a broad range of observations of geology, oceanography, botany, and zoology as described in his *Ansichten der natur* (1808) and *Kosmos* (five vols., 1845). His careful descriptions of the distributions of species over differing latitudes, elevations, and climates established the study of biogeography. Charles Darwin was effusive: "He was the greatest travelling scientist who ever lived." "I have always admired him; now I worship him."

Johannes Eugenius Bülow Warming (1841–1924) was professor of botany at the University of Stockholm (1882–1885), then professor and director of the Botanical Garden in Copenhagen (1885–1911). He identified 370 new species in Brazil. He is recognized as the founder of ecology, closely related to biogeography, *Plante-samfund og haandbog i den systematiske botanik: Grundtræk af den almindelige plantegeografi* (1895).

Botanists had collected and still are collecting masses of data describing the characteristics and distribution of plants on earth. This is interesting in and of itself. Their appropriate classification is essential to communication among fellow botanists and to understanding the history of life. (Chapter C13).

There is an esthetics to collecting and identifying plants; one might regard this as integral to any science. The fact that funding agencies may not consider this justification does not diminish its appeal.

C11

Genetics

Johann Gregor Mendel
(1822–1884).

Darwin got it (mostly) right; he did so without the insights of Mendelian genetics or of contemporary genomics. However, genetics provided the concepts and techniques for the modern synthesis; this is what makes today's biology so exciting. The process and history of evolution is being understood in terms of molecular genetics. Why would the Omnipotent Engineer have designed organisms, cells, and molecules that way? To quote Dobzhansky: "Nothing makes sense in Biology, except in the light of evolution."

Farmers and hunters have selectively bred animals and plants for millennia. Contrary to Darwin's speculation, all dogs descend from wolves (*Canis lupus*), first domesticated about 30,000 years ago. The basic concept is straightforward: breed the individuals, male and female, that have (a tendency toward) the desired characteristics; there was no question that both sexes carry and transmits these traits. A tremendous number of breeds and varietals were developed long before Mendel or the rediscovery of his foundational research. Sometimes engineers can proceed very nicely, thank you, without the benefits of pure research or understanding basic principles.

One might argue that the concepts of Mendelian genetics and the advances in histology and cytology that revealed the chromosome and mechanisms of mitosis and meiosis represented a real paradigm shift. More recent advances in genomics are forcing a re-evaluation of the definition, even the concept, of a gene. However, the legacy of those early plant and animal breeders provided the starting point for genetics.

Jean-Baptiste Lamarck (1744–1829) in *Philosophie zoologique* (1809) proposed the law of use and disuse and the related concept of inheritance of acquired characteristics. He explained that injury or mutilation are not true acquired characteristics and observed that organisms approach perfection, what today would more modestly be called adaptation. Although certain terms — epicycles, alchemy, Lamarckian — are used as pejoratives, Larmack's views of heredity were quite reasonable and were initially embraced by Darwin.

Francis Galton (1822–1911) shared a grandfather, Erasmus Darwin, with Charles. In *Hereditary Genius* (1869), *English Men of Science: Their Nature and Nurture* (1874), *The History of Twins* (1875), and *Inquiries in Human Faculty and its Development* (1883), he argued that intellectual abilities are heritable. This was the opening salvo in the debate, yet to be resolved, of nature vs. nurture.

Johann Gregor Mendel (1822–1884) taught physics at the Augustinian Abbey of St. Thomas in Brünn. From 1856 to 1863 he cultivated and tested ~29,000 peas, *Pisum sativum*, that have easily recognized phenotypes. These *Experiments on Plant Hybridization* were published in the *Proc. Nat. Hist. Soc. of Brünn* in 1866; they were cited only three times over the next 35 years. In 1868 he was appointed abbot of his abbey and turned to more important administrative affairs.

Darwin almost got it. He recorded the numbers of phenotypes and the occasional "sport." Yet by "blending inheritance," the prevailing assumption prior to Mendel, the character of any sport would be diluted away by halves, quarters, eighths... in subsequent generations.

This is a fundamental distinction. In dizygotic organisms, under blending inheritance all of the grandchildren of the Normal/Sport × Normal/Normal matings would be one-quarter sport. Under

Mendelian genetics, three-quarters of the grandchildren would be N/N and one-quarter would be N/S. The sport character would still be observed full strength (N/S), but at low frequency in the population. If N/S provided a selective advantage, if it were adaptive, its frequency in the population would increase over many generations. It is statistically unlikely that the same mutation would occur twice. Hence, with access to "fossil" DNA and to well-dated strata one can deduce the approximate site and time of the original N → S mutation.

Johan Friedrich Miescher (1844–1895) received his M.D. from the University of Basel in 1868. While working in Felix Hoppe-Seyler's laboratory in Tübingen, he isolated nuclein, reasonably pure DNA, from leucocytes in 1871 (Chapter C8). Little did he know...

De Vries, Bateson, and Correns are credited with the re-discovery and extension of Mendel's experiments. Hugo Marie de Vries (1848–1935) held several academic positions — Leiden (1866–1870), Heidelberg (1870), Amsterdam (1871–1875), Halle-Wittenberg (1877), before being appointed professor and director of the Botanical Institute and Garden in 1885. In his *Intracellular pangenesis* he referred to a "*pangene* per trait" in 1889; this was the origin of the word "gene." "*Pangene* per trait" anticipated Garrod's "one gene, one enzyme."

William Bateson (1861–1926) and Punnett described genetic linkage in the *Journal of Genetics* in 1910. Carl Erich Correns (1864–1933) described experiments with the hawkweed plant in *G. Mendel's Law Concerning the Behavior of the Progeny of Racial Hybrids* (1900). He also discussed variegated leaf color, the first description of cytoplasmic inheritance.

Archibald Edward Garrod (1857–1936) studied alkaptonuria, caused by an autosomal recessive mutation of the gene encoding homogentisate 1,2-dioxygenase resulting in the appearance of toxic homogentisic acid (alkapton) in the urine. Alkapton blackens on exposure to oxygen as described in *The Incidence of Alkaptonuria:*

A Study in Chemical Individuality (1902). He coined the phrase "one gene, one enzyme" in *Inborn Errors of Metabolism* (1923).

Thomas Hunt Morgan (1866–1945), professor of experimental zoology, established the famous "fly (*Drosophila melanogaster*) lab" at Columbia University in 1904. With Sturtevant, Bridges, and Muller he published *The Mechanism of Mendelian Heredity* in 1915. He received the Nobel Prize in 1933.

Walter Stanborough Sutton (1877–1916) received his M.D. from Columbia in 1907 and joined the fly lab. He described "reduction division" (meiosis) and wrote: "...the association of paternal and maternal chromosomes in pairs and their subsequent separation during the reduction division...may constitute the physical basis of the Mendelian law of heredity."

George Wells Beadle (1903–1989) received his B.S. in agronomy from the University of Nebraska's College of Agriculture and his Ph.D. in genetics and cytology from Cornell. He worked in the fly lab then accepted the chair of Biology at the California Institute of Technology in 1946. He established that individual mutations affect individual steps in metabolism, an extension of "one gene, one enzyme," for which shared the Nobel Prize in 1958 with George Tatum. He published *The Language of Life* in 1966 while he was president of the University of Chicago (1961–1968).

George Ledyard Stebbins (1906–2000) received his Ph.D. from Harvard in 1931. He explored hybridization and polyploidy in the *Brassicaceae* (parsnips, turnips, kohlrabi, cabbage, Brussels sprouts, cauliflower, broccoli, mustard, rape), *The Significance of Polyploidy in Plant Evolution* in the *American Naturalist* (1940) and *Types of Polyploids: Their Classification and Significance* (1947).

Julian Sorell Huxley (1887–1975) was the grandson of Thomas Huxley, Darwin's "bulldog." He set up the Department of Biology at the newly founded Rice Institute in 1912, was professor of Zoology (1925–1927) at King's College, London, and professor of physiology at the Royal Institution (1927–1931). He was secretary of the Zoological Society of London (1935–1942) and was appointed the first Director-General of the United Nations Educational, Scientific and Cultural Organization (Unesco) in 1946. He was a founding member of the World Wildlife Fund and president of the British Eugenics Society (1959–1962). Of greater, or perhaps lesser importance, depending on one's perspective, he wrote *Evolution: The Modern Synthesis* in 1942, in which he introduced the terms "cline," "ring species," and "clade."

Wright, Fisher, and Haldane are credited with exploring and defining "the modern synthesis" of Darwin's concept of evolution by natural selection and population genetics. Sewall Green Wright (1889–1988) worked in the Animal Husbandry Division of the U.S. Department of Agriculture (1915–1925), then was professor in the Department of Zoology at the University of Chicago (1925–1955). He developed the concepts of inbreeding coefficient, fitness landscapes, genetic drift, and random wandering of small populations. Wright's *Evolution and the Genetics of Populations: Genetics and Biometric Foundations* (four vols., 1969) summarized the modern synthesis over the preceding 50 years.

Ronald Aylmer Fisher (1890–1962) took his B.A. in astronomy from Cambridge University in 1912, then his doctor of science in what would now be called population genetics in 1926. His *The Genetical Theory of Natural Selection* (1930) was foundational. He returned home as a research fellow (1959–1962) with the Commonwealth Scientific and Industrial Research Organisation in Australia.

John Burdon Sanderson Haldane (1892–1964) had a readership in biochemistry, Trinity College, Cambridge (1922–1932), then was appointed professor of genetics at University College, London, in 1932. His *A Mathematical Theory of Natural and Artificial Selection: The Causes of Evolution* (1932) was true to its title. He wrote many articles for the *Daily Worker*, published by the Communist Party of the U.S. He emigrated to India and took a position in the biometry unit of the Indian Statistical Institute in 1953.

Theodosius Dobzhansky (1900–1975) worked in the genetics laboratory at the University of Kiev (1924) before joining Morgan in the fly lab at Columbia; he was supported by a Rockefeller fellowship in 1927. His field work on the natural distribution of alleles, *Genetics and the Origin of Species* in 1937, led to his definition of evolution as "a change in the frequency of an allele within a gene pool." He coined the oft quoted phrase "Nothing in Biology Makes Sense Except in the Light of Evolution."

Barbara McClintock (1902–1992) studied at the Cornell College of Agriculture (B.Sc., 1923, Ph.D., 1927). She took a research position at the Cold Spring Harbor Laboratory in 1942 and served as president of the Genetics Society of America in 1945. She completed the first genetic map of maize in 1931, described the interaction of homologous chromosomes during meiosis, and emphasized the essential role of the tip of the chromosome in ensuring stability to the entire structure. She received the Nobel Prize in 1981 for her work on "mobile genetic elements" or transposons, commonly referred to as "jumping genes." Her initial publications were greeted with some skepticism, not because she was a woman as some would argue, but because her findings forced a fundamental revision of traditional Mendelian genetics.

Frederick Griffith (1879–1941) characterized the polysaccharide capsule of the virulent S strain of *Streptococcus pneumonia* and noted that the avirulent R strain lacked this capsule. Avery, Macleod, and McCarty in 1944 were able to make avirulent strains virulent by exposing them to the "transforming principle" extracted from the S strain. 1952 was a banner year; Hershey and Chase published *Independent Functions of Viral Protein and Nucleic Acid in Growth of Bacteriophage* showing that only ^{32}P labelled DNA, not ^{35}S labelled protein, entered the infected bacterium. These experiments demonstrated that DNA (Meischer's "nuclein") carried genetic information. And Chargaff published his "rule" of base content that $\%A \sim \%T$ and that $\%G \sim \%C$. These experiments, fully as important as the determination of the structure of DNA and with its base complementarity ($A = T$ and $G = C$) and its semi-conservative replication by Watson and Crick in 1953, established modern molecular genetics.

One could hardly understand evolution without insight into the underlying genetics. Yet Darwin got it right, without a real understanding of genetics.

C12

Paleontology

Pterosaur egg

Aristotle and several of his contemporaries recorded their findings of fossils. They were collected for their beauty and uniqueness. Their origins and significance were attributed to various natural and spiritual processes. Hooke in his *Micrographia* (1660) described fossils and speculated as to their origin. Steno suggested that the mineralization of skeletons produced fossils. The early studies of geology by Hutten and others noted consistent patterns of occurrence over large areas of Britain. Cuvier identified *Mosasaur* as an extinct reptile and wrote *Règne animal distribué d'après son organization* (1817). Owen, a bitter rival of Darwin, coined the name dinosaur, terrible reptile, and wrote the definitive *History of British Fossil Reptiles*.

Once it was accepted that these fossils represented the remains of long dead animals and plants, additional concerns presented themselves. Did they still exist in exotic corners of the Earth? If not, why would God permit one of His creations to go extinct? How or why did these strange animals, especially dinosaurs, go extinct? Why were fossils of fish and other aquatic animals found so far above sea level? How could these events be reconciled with Creation and a natural order? Louis Alvarez, Nobel Laureate, 1968, and his son, Walter, proposed in 1980 that a bolide collision 65 million years ago was the source of the iridium found at the K-T boundary and that the resultant global cooling caused the extinction of the dinosaurs. The crater found off the coast of Chicxulub, Mexico, in 1990 is the proposed site of that impact.

The questions posed by this historical science were essential to the development of the theory of evolution. Its more recent findings indicate that the survival of what became *Homo sapiens* was tenuous.

Aristotle (384–322 B.C.) and several of his contemporaries recorded their findings of fossils. They were collected for their beauty and uniqueness. Understanding their origins and significance was essential to the intellectual climate in which Charles Darwin worked. How could God permit one of His creations to go extinct? Surely these animals still thrived in some distant corner of this great Earth. Some fossils closely resembled known animals; how might the immutable have changed? If one accepted gradual evolution by natural selection, why did one not find fossils of the intermediate forms? Is the fossil record not complete? How could the apparent saltation be rationalized? Does the concept and definition of species — animals or plant are members of the same species, if under normal conditions, they can interbreed (elaborated on in Chapter C13) — apply to a slice of time? Is "species," so defined, a valid concept to apply to animals that lived millions of years apart? Is the Popperian concept of falsification, or of verification, applicable to paleontology? If not, how might one evaluate various theories?

Robert Hooke (1635–1703) identified and described fossil shells in *Micrographia*. Nicolas Steno (1638–1686) realized that the commonly found "tongue stones" are actually fossilized shark teeth. Further, the original mineral of fossils (calcium carbonate or calcium phosphate) has been replaced by local minerals over time without change of shape. The details of the several different fossilization processes remain to be elucidated.

Georges Léopold Chrétien Frédéric Dagobert Cuvier (1769–1832) held a range of academic and administrative appointments and wrote extensively about natural history. His *Recherches sur les ossements fossiles* saw three editions (1812, 1821, and 1825). He argued that several catastrophes had led to mass extinctions. Cuvier's theory contrasted with the uniformitarianism of Lyell (Chapter B10) and with Buffon's reluctance to acknowledge extinctions. He was critical of Lamarck's theory of evolution, having seen ancient mummies in Egypt that looked like modern humans — a nice illustration of the importance of understanding the age of the Earth and its species. His reputation is burdened with his "rash dictum" of 1821 that it is unlikely that any large animals remain undiscovered — moral: great to make bold generalizations, but caution advised.

Mary Anning (1799–1847) lived in Lyme Regis on the south coast of England. She followed her father as a hunter of fossils for sale to tourists. In 1821, she found the first complete skeleton of an ichthyosaur, *Plesiosaurus dolichodeirus,* and in 1828, *Pterodactylus macronyx* (renamed *Dimorphodon macronyx* by Richard Owen). She interpreted her extensive findings as evidence for extinction vs. undiscovered living species. She was elected an honorary member of the Geological Society of London; as a woman, she was ineligible for regular membership.

The development of a coherent theory of paleontology depended on the findings of many different people, some of them amateurs. In 1719, William Stukeley described *plesiosaur* as a crocodile, *An Account of the Almost Entire Sceleton of a Large Animal in a Very Hard Stone.* In 1799 Barthélemy Faujas described *Mosasaur* from the chalk quarries of Maastricht as a crocodile. In 1802, Pliny Moody discovered footprints, probably of a theropod, dubbed "Noah's raven." In 1821, William Buckland examined Kirkdale cave in Yorkshire and determined that it was a hyena's den and contained the bones of elephants and other animals, not so common in Britain today; in 1824 he published *Notice on the Megalosaurus,* the first dinosaur described and named. In 1837, Herman von Meyer described *Plateosaurus,* the fifth dinosaur to be named.

Othniel Marsh and Edward Cope livened up the field in the 1870s with their competition to unearth and identify new species in the area of Como Bluff, Wyoming. Their combined total was 135. With each passing decade over two centuries there have been ever more fascinating finds, not only of dinosaurs but of all sorts of extinct animals and plants. In 1877, Paul Gervais presented thin section micrographs of fossil dinosaur eggs. In 1888, Harry Seeley distinguished "lizard-hipped" (saurischian) from "bird-hipped" (ornithischian) dinosaurs.

Richard Owen (1804–1892) received his M.D. from the University of Edinburgh in 1824; he was appointed Hunterian professor of the Royal College of Surgeons in 1836 and served as conservator until 1856, when he became superintendent of natural history of the British Museum. He wrote *Comparative Anatomy and Physiology of Vertebrates* (3 vols., 1866–1868), *History of British Fossil Reptiles* (4 vols., 1849–1884), and *Catalogue of the Fossil Reptilia of South Africa* (1876). He coined the term "dinosaur" (terrible lizard). He was hardly a vitalist but spoke of living matter having an "organizing energy." In 1860, he wrote an anonymous review of Darwin's *Origin* for the Edinburgh Review complaining that Darwin had ignored his "axiom of the continuous operation of the ordained becoming of living things." Darwin said: "The Londoners say he is mad with envy because my book is so talked about. It is painful to be hated in the intense degree with which Owen hates me." "I used to be ashamed of hating him so much, but now I will carefully cherish my hatred and contempt to the last days of my life."

In 1922, Roy Andrews from the American Museum of Natural History went to Mongolia in search of early hominids and found massive beds of dinosaurs. In 1971, Aleksandr Sharov identified prints of hair on *Pterosaur* in Kazakhstan, and Douglas Lawson unearthed *Quetzalcoatlus northropi* with a 12 m wingspan in Texas. In 1996, Chen Pei Ji described the first feathered *Sinosauropteryx*. Some 500 genera of dinosaurs have been identified and their relationships and lifestyles convincingly postulated.

Luis W. Alvarez (1911–1988) received his B.A. (1932) and his Ph.D. (1936) from the University of Chicago. During the Second World War, he worked on exploding bridge wire detonators for the Manhattan Project. He was awarded the Nobel Prize in 1968 for the "discovery of a large number of resonance states, made possible through his development of the technique of using hydrogen bubble chamber and data analysis." In 1980, he and his son, Walter, published the asteroid impact theory based on the finding of elevated amounts of iridium, rare in the earth's crust but more abundant in asteroids, at the K-T (cretaceous-tertiary) boundary. They argued that the dust thrown into the atmosphere blocked sunlight with the resultant cooling of the Earth and extinction of the dinosaurs and many other species of animals and plants. In 1990, a huge crater was found in the Gulf of Mexico near Chicxulub that was dated to the K-T and is generally agreed to fit Alvarez's prediction. This provides an example of verification, if not quite attempted falsification, after Popper, in a historical science (Chapter A10).

Homo sapiens survived the release of *Jurassic Park* in 1993. Human bones also fossilize with time; in the short term they contain fragments of DNA. Archeology and physical anthropology are, or should be, closely related.

In 1868, Ernst Haeckel published *Natürliche Schöpfungsgeschichte* (*History of Creation*), in which he proposed 12 human species and 22 phases of human evolution. The twenty-first was the "missing link" between apes and humans. In 1826, William Buckland discovered the first human fossil, so-called "Red Lady," a Cro-Magnon ("big hole") male, 33,000 ybp, on the Gower Peninsula, Wales. In 1856, the first recognized *Homo neanderthalensis* was unearthed near Düsseldorf. In 1863, Julian Huxley published *Evidence as to Man's Place in Nature*. In 1868, three more skulls of *Cro-Magnon*, dated about 30,000 ybp, were found. In 1894, Eugene Dubois discovered *Pithecanthropus erectus*, Java Man, designated as the missing link.

In 1896, Cunningham first described Neanderthals. In 1907, *Homo heidelbergensis* was suggested to be the precursor of Neanderthals. In 1909, Henri Breuil found in France skeletons of Neanderthals who had been buried. In 1911, a Neanderthal hand axe, ~200,000 ybp, was found near Norfolk. In 1911, Charles Dawson found the "Piltdown skull" in England. It was exposed as a fake in 1953; however, the villain remains a target of speculation. In 1921, miners in Rhodesia found skull and legs of *Homo heidelbergensis*. In 1923, Teilhard de Chardin found ancient stone tools from the Paleolithic in Shuidonggou.

Raymond Dart in 1925 found in Africa a skull called "Taung Child"; it was named *Australopithecus africanus*. In 1929, Davidson Black found "Peking Man" in Zhoukoudian, *Sinanthropus pekinensis*. In 1948, Mary Leakey found the skeleton of the ape "Proconsul" dated $16 \cdot 10^6$ ybp. In 1948, Donald Johanson discovered another *Australopithecus afarensis* (Lucy) and presented evidence that she was bipedal. In 1949, A. Leroi-Gourhan excavated the Grotte du Renne (Cave of the Reindeer) at Arcy-sur-Cure; he found shells collected by its Neanderthal inhabitants. In 1959, Mary Leakey unearthed a skull of *Australopithecus boisei*. In 1964, Louis Leakey found *Homo habilis*. From 1967–1974, Richard Leakey, son of Mary and Louis, unearthed the skulls "Omo I" and "Omo II," *Homo sapiens*, in Ethiopia, dated 195,000 ybp. In 1978, Mary Leakey found "fossil" footprints of a bipedal at Laetoli, dated $3.6 \cdot 10^6$ ybp. In 1984, Richard Leakey found Turkana Boy, *Homo erectus*. In 1987, Allan Wilson and his students proposed, based on the sequences of mitochondrial DNAs, that all humans "out of Africa" share a common female ancestor, subsequently nicknamed "mitochondrial Eve," from about 150,000 ybp.

One might imagine that the Earth had been scoured clean but the pace of fossil finds increases.

In 2002 David Lordkipanidze found a skull of *Homo erectus*, "toothless old man," in Dmanisi, Georgia, dated $18 \cdot 10^6$ ybp. In 2002, Michel Brunet found *Sahelanthropus tchadensis* in Chad, $\sim 6 \cdot 10^6$ ybp. In 2004, Naama Goren-Inbar presented evidence of controlled fire used by hominids in Israel, 790,000 ybp.

In 2004, Peter Brown and Mike Morwood described skeletons of *Homo floriensis*, nick-named Hobbit, from a cave in Flores, Indonesia. Hobbits, dated as recent as 18,000 ypb, were

1.0 m tall and had brain volume of 380 vs. 1400 cm^3 of *Homo sapiens*. In 2005, Haile-Selassie *et al.* found a nearly complete skeleton of *Australopithecus anamensis* in Ethiopia, $4 \cdot 10^6$ ybp. In 2006, Zeresenay Alemseged summarized the evidence that *Australopithecus afarensis* was bipedal. In 2007, F. Spoor and Meave Leakey found fossils of *Homo habilis* and of *Homo erectus* near one another, in the same strata, 500,000 ybp, strongly implying their co-existence. In 2009, Christopher Henshilwood found 13 engraved ochre (Fe_2O_3 stone with clay admixture) artifacts in the Blombos cave, South Africa, 100,000 ybp. In 2010, DNA was extracted from remnants of flesh on the finger of a *Homo denisovan* found in a cave in Siberia, 48,000–30,000 ybp. In 2010, fossils of *Australopithecus sedia* were found near Johannesburg, South Africa, $2 \cdot 10^6$ ybp.

The early findings of paleontology contributed to Darwin's thinking and conversely the concepts of evolution and speciation were essential to understanding these discoveries.

Frequently the soils of jungles are warm, moist, and acidic, hence the rapid degradation of the both flesh and bone. Few fossils have been found of the great apes, *Pan* and *Gorilla*. As *Homo* became bipedal and left the trees for the savannah, he also left more fossils. The record of the evolution of *Homo sapiens*, and of its extinct con-specifics, over the past five million years is remarkably complete and certainly the pace of discovery will continue. In part this reflects the interest in and funding for such studies as well as the sophisticated methods of analysis and dating. Only in the last $\sim$18,000 years have we, *H. sapiens*, been the only hominid on our planet. Svante Pääbo summarized the evidence that anatomically modern humans overlapped and mated with Neanderthals such that non-African humans inherit $\sim$1 to 3% of their genomes from Neanderthal ancestors.

Given the extinction of so many hominids, the survival of *H. sapiens* might have seemed tenuous. Now numbering >7,000,000,000 and growing, one might again wonder how tenuous?

C13

Systematics

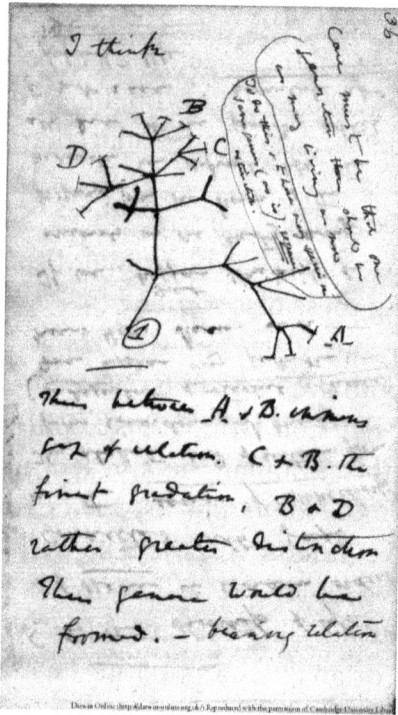

From 'B' notebook
(Charles Darwin, 1837).

Some of those engaged in the early study of natural history were academicians; probably the majority were amateurs. Well before 1800, many authors assembled encyclopedic works describing all of the known animals and plants — their morphologies, habitats, and behaviors. How to name them? How to organize this ever growing trove? Linnaeus is properly honored for having introduced and standardized a binomial system, genus and species with Latin names. But this begs the questions: How to define or characterize a genus or a species? What is the proper relationship(s) among genera?

Implicit in these questions was the concept and the search for a natural system, or better *the* natural system appropriate to plants and to animals, monocots and dicots, vertebrates and invertebrates.

One already had the *Scala Naturae* (Great Chain of Being) of Aristotle; though not explicitly rejected, its inappropriateness became ever more obvious.

If the natural system is hierarchical, might it reflect the actual process of evolution? The development of this natural system is absolutely non-trivial and represents one of the great achievements of biology. Yet how did such a scheme come to be? Why should the Creator be so constrained?

The natural system sought by these biologists is quite different from the ordering of the elements by Newlands and Mendeleev. Further, these two "natural orders" — one of biology, the other of chemistry — posed quite different questions and yielded very different, but extremely important, insights into the workings of nature.

It is gratifying to see that hierarchies based on comparisons of DNA sequences match quite closely those based on traditional traits. Taxonomy and cladistics are still vital fields with hotly contested questions: How best to construct the branching points and branch lengths of dendrograms given various observations? How to weight these observations? Given that all branches do not line up at similar levels corresponding to genus, to family, to order, etc., how to rank and name the branches? With the passage of time and speciation, ought one extend the species name back beyond the branch point? Surely our present system of genus and species is, in this sense, not natural though universally accepted.

Can this binomial system be applied to eubacteria and archaea? To viruses? Given the frequency with which horizontal gene transfer occurs among bacteria, is a hierarchical classification applicable to all life on Earth?

As is often the case, once one knows the answer the question is easy. Given a lot of observations or objects, how does one classify them? Even naming the entities explicitly or implicitly implies some ordering. The challenge in biology took several forms. Although seldom elaborated, one basic observation is that there is not a continuum of animals and of plants. One sees either wolves, or coyotes, or foxes; but not (usually) intermediates. For openers one can assign the term "wolf" to certain beasts and not to others. If one encounters a new carnivore, how can one, aside from subjective intuition, decide to call it a wolf, yea or nay? Is there an essential feature, or "eide," that makes this new specimen a wolf or not? Or, does "wolfness" encompass a group of characters and "foxness" a different group — 7 of 10, it's a wolf; 6, not? Perhaps it doesn't score 7/10 for foxness either; one may then create a new group, coyote. Such abstractions might seem unreal and unnecessary to a field biologist whose intuition is well honed and is usually right. It took several centuries for the community to accept the definition summarized by Ernst Mayer (1904–2005)" "Two organisms are members of the same species, if under natural conditions they would interbreed" (no ligers or tigons from zoos).

Given that most animals and plants can be assigned to a species, how to order or find a relationship among the distinct species? Aristotle had proposed a hierarchical scheme, *Scala Nature* (Great Chain of Being), in which animals were above plants, and humans above animals. As will be discussed in Chapter C15, this approach, based on "values," readily lends itself to racism, with the home team inevitably on top.

If one accepts identifiable groups, e.g. species, one might order them from top down by repeated divisions — plants vs. animals; vertebrate vs. non-vertebrate; viviparous vs. oviparous — ever downward. Or, one might work from the bottom up, clustering observed characters. In practice, most naturalists use both approaches without explicitly acknowledging or endorsing one or the other.

Should one use only (skeletal) morphology, as available to paleontologists, or might one also consider physiology, development, behavior, and ecology? If one uses this full range should they be weighted; if so, how? As one goes up or goes down should one use the same characters and weights? It isn't very meaningful to ask whether a plant has vertebra. These biological classifications are not in practice algorithmic; however, in retrospect they may be described as rational. Many, but certainly not all, would argue that DNA sequences are the gold standard and should take precedence. The community breathed a collective sigh of relief when it was found that dendrograms based on DNA are quite similar to those based on traditional criteria, e.g. *The Scientific Basis of Classification*, Williams and Norgate, 1882.

As noted, the history of pure science, as opposed to applied — engineering and medicine — is best told in terms of human beings, as opposed to paradigms or programs or schools. Yet these people, consciously or not, do work within the context of a prevailing ethos.

Theophrastus (370-285 B.C.) is regarded as the "father of taxonomy" after his groupings in *Enquiry into Plants* and *On the Causes of Plants*. He was critical of Aristotle's "Great Chain of Being" as a guide to classification. Pliny the Elder (23–79) mixed natural history, fantasy, and moralizing in his compilations. Dioscorides (~40–~90) listed 500 plants in his pharmacopeia, *De materia medica*, by medicinal application.

Otto Brunfels (1488–1534) ordered his entries alphabetically in *Herbarum vivae icons* (1530–1536) in three parts and in *Contrafayt Kräuterbuch* (1532 and 1537) in two parts. Hieronymus Bock (1498–1554) listed 700 species in his *Kreuterbuch* (1539). Konrad Gessner (1516–1565) wrote *Enchiridion historiae plantarum* (1541), *Catalogus plantarum* (1542),

Historiae animalium (1551–1558); these were the best compilations of life, as known at the time.

Many were seeking a "natural system" of classification without an explicit statement of how it might be characterized or recognized. Andrea Cesalpino (1519–1603) was director of the botanical garden of Pisa (1554–1558). His *De plantis libri XVI* (1583) was the first systematic classification of flowering plants; it was based on the structures of their reproductive organs. John Ray (1627–1705) wrote several catalogs of plants and animals culminating in his *Historia plantarum* (1686–1704), which listed 18,600 plants. He advocated the use of several unweighted characters in classification. John Locke (1632–1704) questioned whether the concept of species was made by man or whether it was natural: "I think it nevertheless true that the boundaries of species, whereby men sort them, are made by men . . ."

Carl Linnaeus (1707–1778) authored 10 editions of *Systema naturae* from 1735 to 1758; the 10th contained 7,700 species of plants and 4,400 animals. He introduced the binomial naming system — genus, species — still used today. He was skeptical of either extinction or evolution. "We count as many species as different forms were counted in the beginning." Michel Adanson (1727–1806) worked in Senegal (1749–1753) as an employee of the Compagnie des Indes. On returning to Paris he had access to the French royal collection of plants. His *Les familles naturelles des plantes* (1763) sought a natural system of classification and nomenclature of plants, based on all of their physical characteristics. He laid the foundations of numerical taxonomy. Ernst Heinrich Philipp August Haeckel (1834–1919) traveled in Dalmatia, Egypt, Greece, Norway, Turkey and the Canary Islands. He introduced the terms "phylum" and "phylogeny" as well as the kingdom *Protista*, now characterized as single cell eukaryotes with a nuclear membrane. His dictum "ontogeny recapitulates phylogeny" is not now used as a basis of classification; however, it is generally valid (Chapter C5).

Johannes Eugenius Bülow Warming (1841–1924) was professor of botany at the University of Copenhagen and director of the Botanical Garden (1885–1911). He was one of the founders of ecology and considered geographic distribution to be a significant characteristic in the definition of species, *Haandbog i den systematiske Botanik* (*Handbook of Systematic Botany*, 1878).

Georges-Louis Leclerc, Comte de Buffon (1707–1788), wrote *Histoire naturelle* not for identification or for classification, but for appeal to the general reader. Lamarck (1744–1829) proposed branching "trees," like contemporary dendrograms, to display relationships among species.

Georges Cuvier (1769–1832) proposed phyla of vertebrates and invertebrates; within invertebrates were mollusks, crustaceans, insects, worms, echinoderms, and zoophytes (*Mémoires pour servir à l'histoire et à l'anatomie des mollusques*, 1817). Kant (1724–1804) was more insightful: "The natural classification deals with lines of descent, grouping animals according to blood kinship."

Darwin (1809–1882) considered classification essential:- "...propinquity of descent — the only known cause of the similarity of organic things — is the bond, hidden as it is by various degrees of modification which is partially unveiled to us by our classifications." He considered the degree of divergence to be reflected in the proximity of branch points:

> All true (biological) classification is genealogic...that the arrangement of the groups within each class, in due to subordination and relation to the other groups, must be strictly genealogical in order to be natural...that the less any part of the organization is concerned with special habits, the more important it becomes for classification." — *Origin* (p. 420).

In contrast to Cuvier, he felt that the value of a character for classification was inversely proportional to its value for survival of the organism.

Nigel Goldenfeld and Carl Woese in *Biology's Next Revolution* (2007) argued that "...microbes as gene-swapping collectives demand a revision of such concepts as organism, species and evolution itself." "... horizontal gene transfer, the non-genealogical transfer of genetic material from one organism to another... microbial behavior must be understood as predominantly cooperative." "Rather than discrete genomes, we see a continuum of genomic possibilities... Refinement through the horizontal sharing of genetic innovations would have triggered an explosion of genetic novelty, until the level of complexity required a transition to the current era of vertical evolution."

Horizontal gene transfer (HGT) is extremely rare among eukaryotes, especially plants and animals. Species are very well differentiated in animals in which reproductive tissues are set aside early in development (Chapter C5). In contrast HGT is common in bacteria and could lead to "a continuum of genomic possibilities." If so, why are there distinct genera and species, e.g. *Stapholycoccus* and *Streptococcus* instead of a continuum? What better proof of a malevolent Divine than to let us think that we understood — then say gotcha?

C14

Evolution

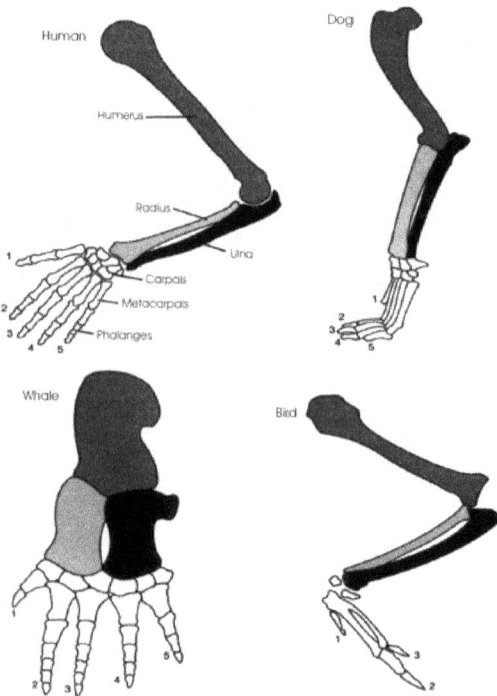

The bones in the forelimbs of terrestrial and some aquatic vertebrates are remarkably similar because they have all evolved from the forelimbs of a common ancestor.

Darwin, along with Newton, Maxwell, and Einstein, is rightly considered to be one of the greatest scientists. Certainly his concept of evolution by natural selection provides the core, conceptual framework for biology. The goal of many of these chapters has been to summarize the status of the several disciplines of physics and of biology up to 1859, as Darwin was formulating his argument, then on to the early 1900s, by which time most of its tenets had been accepted. As one of those lovely coincidences (or was it a coincidence?), this was also the dawn of modern physics as well as of modernity in the arts.

Darwin, from the beginning of the second voyage of the Beagle (1831–1836) until the publication of the first edition of the *Origin of Species* (1859), did not have access to any privileged information. He was able to sort through many current discussions, as well as his own personal observations and collections. With advantage of hindsight one might summarize the information and intellectual climate available to him:

1) The Earth was old; it was very old. There was time enough for tiny, incremental changes to accumulate and make a major change in a species.
2) The various members of a species were not identical to one another; there was not an "essential" character for each species.
3) A "natural" system of classification was hierarchical and accommodated the observations of inferred speciation and extinction.
4) Some members of a species begat more viable offspring than did others.
5) The various members of a species can interbreed; members of different species do not interbreed.
6) The scientific revolution (1600–1700) had established an intellectual climate of analysis.
7) Most scientists professed their faith; the Church remained sensitive to challenges to its doctrine.
8) Morphology, especially skeletons of fossils, showed definite similarities, homologies, among species.
9) Early embryos of vertebrates are quite similar; differences characteristic of a species appear later in embryonic development.
10) Malthus had very persuasively argued that the number of organisms in a species would increase geometrically, if unchecked. Averaged over decades the population size of a species remains approximately constant.

Although the broad concept of evolution was soon accepted by the general scientific community, if not by the Church(es), three-quarters of a century passed before there was agreement on mechanisms and implications. In Chapter D4, we consider some non-accepters a century and a half later.

No aspect of science, including the shift from helio-centrism to geo-centrism, has so threatened Christian beliefs as has the general concept of evolution by natural selection and the specific idea of *Homo*'s sharing a common ancestor with *Pan* and *Gorilla*. How can this be reconciled with *Genesis 1.1* "In the beginning God created the heaven and the earth. And the earth was without form, and void; and darkness was upon the face of the deep. And the Spirit of God moved upon the face of the waters. And God said, Let there be light: and there was light." *Genesis, 1: 26:* "And God said, Let us make man in our image, after our likeness: and let them have dominion over the fish of the sea, and over the fowl of the air, and over the cattle, and over all the earth and over every creeping thing that creepeth upon the earth."

The intellectual climate in which Charles Darwin (1809–1882) and Alfred Wallace (1823–1913) developed their ideas can be summarized in 10 broad concepts: time, essentialism, classification, selection, species, scientific revolution, faith, embryology, and population.

1) Time: During the Renaissance there was no general understanding of the age of the Earth; it was young. James Ussher in 1648 published his calculation, based on careful interpretation of scriptures, that the Earth was created on 23 October, 4004 B.C. Nicolas Steno in 1669 identified geological strata. In 1779, Buffon, author of the encyclopedic *Histoire Naturelle*, estimated that the earth was 75,000 years old. James Hutton, about 1790, thought that the Earth was old and Charles Lyell thought it was very old, with perhaps no beginning and no end. Lord Kelvin in 1869 calculated, quite reasonably, the age of the earth to be 3×10^7 years, based on heat transfer from a molten sphere of iron (Chapter B10); how might he have anticipated radioactive decay?

2) Essentialism: Plato thought in terms of fixed geometries, a harmonious cosmos and time invariance. Democritus postulated building blocks, atoms, of constant character. In contrast Heraclitus described a flux in nature and Aristotle referred to "adaptation" of organisms to their environments and to a *scala natura* as a foundation of natural history. Gottfried Leibniz argued that every possible organism had been created and that appearance might change but not a perfect internal essence (Chapter A2). By 1850, essentialism, or typological thinking, was on the wane but still a topic of discussion. Within a species there is variation.

3) Classification: The concept of distinct species and their assignments to genus by Linnaeus was of fundamental value, as attested by our continued use of his scheme nearly 200 years later. The greater challenge was how to order these genera. John Ray in *Historia plantae* (1686) had argued:

> In order that an inventory of plants may be began and a classification of them correctly established, we must try to discover criteria of some sort for distinguishing what are called "species". After a long and considerable investigation, no surer criterion for determining species has occurred to me than the distinguishing features that perpetuate themselves in propagation from seed. Thus, no matter what variations occur in the individuals or the species, if they spring from the seed of one and the same plant, they are accidental variations and not such as to distinguish a species... Animals likewise that differ specifically preserve their distinct species permanently; one species never springs from the seed of another nor *vice versa*.

Several questions were addressed — single vs. multiple characters? Should ecological or behavioral characters be used in addition to morphology? Should multiple characters be weighted and if so how? There was a sense of a "natural classification"; yet, no one could articulate exactly what it was (Chapter C13). Darwin wrote:

> In short, we shall have to treat species in the same manner as those naturalists treat genera, who admit that genera are

merely artificial combinations made for convenience. This may not be a cheering prospect; but we shall be freed from the vain search for the undiscovered and the undiscoverable essence of the term species. — *Origin*, p. 282.

He concluded that "all true classification is genealogical." As more fossils were unearthed, and as unexplored regions of the Earth decreased, the concept of plentitude, as argued by Leibniz, became ever more tenuous. Further, these fossils were found throughout the world in the same strata (Chapter B10). The creation and extinction of species is an ongoing process on this very old Earth. One of the great challenges to naturalists was (and is) to refrain from assigning human values, such as Aristotle's *scala natura*, or "Great Chain of Being" to natural phenomena (Chapter A2). It seemed irresistible to popularizers of science and to authors of children's stories, in all cultures, to assign virtue, or lack thereof, to chosen animals, be it Bacon's industrious bees or the wily coyote.

4) Selection: The description(s) of inheritance of acquired characteristics by Jean-Baptiste Lamarck was quite reasonable and innovative when viewed in context. The distribution of characteristics within a species may vary with environment. He extended this observation and argued that "one species transformed into another," and acknowledged the possibility of the elimination of species. It is ahistorical to dismiss epicycles (Chapter B3), alchemy (Chapter B5), or Lamarckism (Chapter C13). Darwin was influenced by the arguments of Lamarck and included "soft" selection in early editions of *Origin*.

5) Species: Darwin wrestled with the concept of species: "To sum up, I believe that species come to be relatively well-defined objects, and do not at any one period present an inextricable chaos of varying and intermediate links." (*Origin*, p. 177). It became ever more accepted that species are the fundamental unit of biology, the coin of the realm. The generally accepted definition was best articulated by Dhobzansky, to paraphrase:

"All members of a species can interbreed with one another; under normal conditions members of different species to not interbreed." The generation of hybrids, especially in plants, complicates but does not invalidate this definition.

6) Scientific Revolution: The scientific revolution and subsequent industrial revolution had set an atmosphere of critical analyses. One could no longer resort to mythos to explain natural phenomena.

7) Faith: Most scientists professed their faith. Many argued that understanding how His creations worked deepened their appreciation. Nonetheless, the Church remained sensitive to challenges to its doctrine. The immolation of Bruno and the trial of Galileo (Chapter B4) were still in mind. Darwin was well aware that the Church would not be pleased that *Homo* and *Pan* shared a common ancestry.

8) Morphology: Morphology was easily observed and seemed fundamental. Darwin noted: "This is the most interesting department of natural history and may be said to be its very soul." (*Origin*, p. 434). It seemed a solid metric and could be applied to fossilized skeletons.

9) Embryology: Darwin continued: ". . . most recent evolutionary acquisitions are due to variations that had occurred very late in ontogeny." "Thus, community in embryonic structure reveals community of descent." Haeckel argued that "ontogeny follows phylogeny." That is, early embryos of vertebrates are quite similar; the differences that characterize species appear later in embryonic development (Chapter C5). This observation informs models of both classification and selection.

10) Population: Thomas Malthus (1766–1834), acknowledging the earlier comments of Benjamin Franklin, wrote: "It may safely be pronounced, therefore, that population, when unchecked, goes on doubling itself every 25 years, or increases in a geometrical ratio." Yet, populations are approximately stable over many generations. He concluded that there is a "struggle for existence." Seemingly trivial observations can have profound implications.

Darwin's interests and views changed with time; they evolved. He acknowledged the struggle for existence among individuals of a population. Survival is not random; it depends, in part, on heritable traits of survivors. The gradual change in frequency of these traits within populations leads to new species. He wrote: "As I have said, *isolated* species, especially with some change (in environment) probably vary quicker." This term "isolated" captures the essence of allopatric speciation. He was very familiar with the similar process in the artificial selection practiced by plant and animal breeders.

Although today we recognize evolution by natural selection as the core idea of biology, it was not fully accepted by the general community until the "modern synthesis" of the 1930s. Nearly all biologists accepted *evolution*; however, views of its mechanism(s) and implications varied. In simplification (Ernst Mayr, 1904–2005):

	Common Descent	Gradual Change in Population	Speciation	Natural Selection
Lamarck	No	Yes	No	No
Darwin	**Yes**	**Yes**	**Yes**	**Yes**
Haeckel	Yes	Yes	??	??
neo-Larmarck	Yes	Yes	Yes	No
Huxley, T.	Yes	No	No	?No
deVries	Yes	No	No	No
Morgan	Yes	?No	No	??

Contrary to some popular simplifications, Darwin did not address the origin of life (Chapter C17) in his publications. He did write a letter to Joseph Hooker in 1871: "But if (and oh what a big if) we could conceive in some warm little pond with all sorts of ammonia and phosphoric salts, light, heat, electricity *et cetera* present, that a protein compound was chemically formed, ready to undergo still more complex changes..."

C15

Race

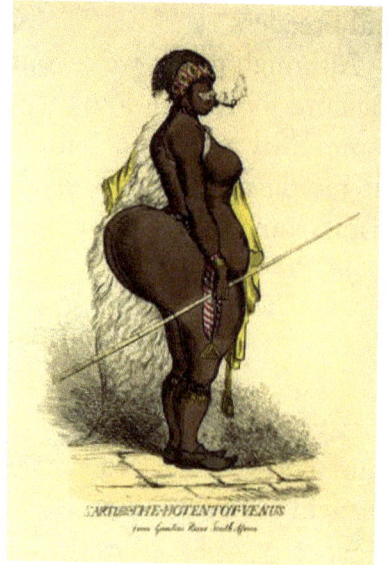

Saartjie Baartman-Hottentot Venus (Khosi, ~1805).

Groups of people — families, tribes, nations — have often treated other groups with suspicion and hostility. Whether this behavior is genetically encoded and/or adaptive is debated. The "other" can be characterized by appearance, custom, or speech. Many societies — Egypt, China, and Greece — described other races and came to the inevitable conclusion of their own superiority. This hierarchical ranking is consistent with Aristotle's "Great Chain of Being."

Many societies have held people in bondage; the *New Testament*, Ephesians (6:7-5) admonished:

> Slaves, be obedient to the men who are called your masters
> in the world, with deep respect and sincere loyalty as you are
> obedient to Christ: not only when you are under their eye, as
> if you only had to please men, but because you are slaves of
> Christ and wholeheartedly do the will of God...

Slavery was hardly invented in the sugar plantations of Latin America or the cotton fields of the American South. Yet the scale of these operations and the ready supply of slaves from West Africa led to a system of slavery seldom seen before. It was much easier to condone slavery if the enslaved were further down on the Great Chain.

The movement(s) to abolish the slave trade, then slavery, coincided with a legitimate curiosity about non-European peoples. The subsequent studies of race and of eugenics were not inherently evil but were easily exploited. We are still wrestling with the challenges of racial equality and the legitimate characterization of genetic differences among individuals as well as among identifiable groups. These concerns beg the question of human origins, evolution, and diversity.

Some of the more significant findings of hominoid fossils or remains are listed (from Chapter C12). The pace of discovery and sophistication of analysis quickens; stay tuned.

	Genus species	Site or name	Discoverer
1856	*Homo neanderthalensis*	Engis	Philippe-Charles Schmerling
1891	*Homo erectus*	*Java man*	Eugene Dubois
1924	*Australopithecus africanus*	*Taung child*	Raymond Dart
1929	*Homo erectus*	*Peking man*	Davidson Black
1959	*Parathropus boisei*	Tanzania	Mary Leaky
1960	*Homo habilis*	Tanzania	Mary and Louis Leaky
1992	*Ardipithecus ramidus*	Ethiopia	Tim White
1994	*Ardipithecus ramidus*	Ethiopia, *Ardi*	J. Haile-Selassie
1994	*Australopithecus anamensis*	Kenya	Meave Leaky
1997	*Ardipithecus kadabbai*	Ethiopia	J. Haile-Selassie
2000	*Orrorin tugenensis*	Kenya	Martin Pickford
2002	*Sahelanthropus tchadensis*	Chad	Michel Brunet
2003	*Homo floresiensis*	*Flores Man*	Mike Morwood
2005	*Homo denisova*	Siberia	Reich *et al.*

The earliest known examples of several "species" of hominid are listed (and in parentheses youngest fossil).

Millions years before present

6.7 *Sahelanthropus tchadensis*
6.1 *Orrorin tugenensis*
5.7 *Ardipithecus kadabba*
4.4 *Ardipithecus ramidus*
4.0 *Australopithecus anamensis*
3.4 *Australopithecus afarensis (Lucy)*
2.4 *Australopithecus africanus*
2.5 *Paranthropus aethiopicus*
1.8 *Paranthropus boisei*
1.9 *Homo habilis*
1.3 *Homo erectus*
0.5 *Homo heidelbergensis*
0.3 *Homo neanderthalensis* (to 35,000 ybp)
0.3 *Homo denisova* (to 30,000 ybp)
0.2 *Homo sapiens*
0.1 *Homo floresiensis* (to 20,000 ybp)

Homo sapiens overlapped with *neanderthalensis, denisova,* and *floresiensis* temporally and quite possibly geographically. Whether they interacted in hunting, in combat, or in amour is the subject of DNA analysis and intense speculation (Pääbo, *Neanderthal*).

The definition of species for extant organisms is generally accepted. "Member of a species can interbreed; members of different species cannot interbreed under normal circumstances." How can one extend this definition to extinct species or to fossils? If the bones are quite similar (and that begs the question) and if the fossils are found in similar strata, it is reasonable to assign them to the same species. However, folks who resided in caves cannot be easily stratified. A professional career is more enhanced by naming a new species than by discovering another representative of a named species.

Present models indicate that *Homo habilis* occupied the savannahs of East Africa 2,000,000 ybp. *Homo erectus* crossed into the Middle East about 1,700,000 ybp and migrated throughout the Old World. *H. erectus* went extinct; when, where, why remains beyond speculation.

Homo sapiens is also inferred to have evolved from *Homo habilis*, as recently as 200,000 ybp. In the generally accepted "out of Africa" model(s) there were (at least) two distinguishable out migrations of *H. sapiens*. The first occurred about 60,000 ybp; the only remaining "pure" or non-interbreed representatives today are the aboriginals of Australia. The second out migration occurred about 30,000 ybp. The origins and migrations of *H. neanderthalensis*, *H. denisova*, and *H. floresiensis* are less well defined, Reich *et al.* (2011).

Consistent with this model the diversity within Africans is greater than that among all Eurasians. "Sequencing of Bushmen [Khoisan] genomes shows that they have more genetic differences with one another than a European would with, say, an Asian," Rasmussen *et al.* (2011).

A fixation index, F_{ST}, of 0.00 indicates complete panmixis; 1.00 is complete separation. $F_{ST} > 0.25$ usually warrants the designation of "subspecies." Humans ~0.15; leopards ~0.24; grey wolves ~0.90. Based on the diversity of sequences of mitochondrial DNA of Eurasians and on reasonable rates of fixation of random mutations, Wilson *et al.* (1985) suggested that the mitochondrial Eve(s) came out of Africa about 20,000 ybp. A similar "out of Africa" date has been derived for Eurasian males based on Y chromosome haplotypes. Mendez *et al.* (2013) did a similar analysis of DNA from the Y chromosomes and suggested a date of ~300,000 ybp for the Y-Adam. The estimated date for the mitochondrial-Eve ~175,000 ybp. The two dates are not expected to be the same.

"The average proportion of nucleotide differences between a randomly chosen pair of humans (i.e., average nucleotide diversity) is consistently estimated to lie between 1 in 1,000 and 1 in 1,500.

This proportion is low compared with those of many other species, from fruit flies to chimpanzees," Ebersberger *et al.* (2002); Yu *et al.* (2003); Fischer *et al.* (2004).

Bauchet *et al.* (2007) examined 10,000 DNA polymorphisms and identified, even with this low level of diversity, five clusters within Europe — south-eastern, northern, Basque, Finnish, and Spanish. The human Genome Diversity Project documented ~100,000 single nucleotide polymorphisms (SNPs) and differences in copy number of genes from 50 geographic groups. They confirmed that populations of *H. sapiens* lost genetic variation as they migrated from Africa and that it is possible to trace an individual's geographic heritage via his/her SNPs. They concluded: "A huge amount of our genomes are the same across the world, and that helps to argue against racism..."

The ancients considered race to be important and didn't wait for DNA analyses. The Egyptian *Book of Gates* recognized Egyptians, Asiatics, Libyans, and Nubians. The three sons of the biblical Noah — Shem, Ham, and Japheth — founded the Semitic, African, and Eurasian races.

Prejudices of religion and of race were intertwined. Robert Boyle (1627–1691), as director of East India Company, promoted Christianity in the East and founded the Boyle lectures "...to defend Christianity against notorious infidels, namely atheists, deists, pagans, Jews and Muslims." Carl Linnaeus (1707–1778) considered the species of man to be *Homo africanus, americanus, asiaticus, europeanus,* and *monstrosus.* Georges-Louis Leclerc Comte de Buffon (1707–1788) accepted a common ancestry of humans and apes. He considered the creatures of the New World inferior to their counterparts in Eurasia; their peoples are less virile, perhaps due to the marsh odors and dense forests of America. Buffon was a gifted popularizer of science but was not as original a thinker as was Bernard Le Bovier de Fontenelle (1657–1757), who published *Entretiens sur la pluralité des mondes (The Plurality of Worlds)* in 1686.

James Cowles Prichard (1786–1848) published five volumes of *Researches into the Physical History of Man* from 1813 to 1847. He argued for the primitive unity of the human species "...acted upon by causes which have since divided it into permanent varieties or races." "On the whole there are many reasons which lead us to the conclusion that the primitive stock of men were probably Negroes, and I know of no argument to be set on the other side." He was a member of the Aboriginal Protection Society and supported the Aboriginal Protection Act of 1869.

In contrast Jean Louis Rodolphe Agassiz (1807–1873) championed polygenism; that is the different races had separate origins, as created and distributed by God (*Diversity of Origin of the Human Races*, 1850). Thoughtful people came to quite different conclusions. The burden of proof rests with those who would ascribe negative racism, subconscious or otherwise, to their analyses.

Wilberforce (1759–1833) argued that Britain must reject "...the wild and guilty phantasy that man can hold property in man." In 1807, Britain and the United States outlawed the international transport of slaves. Brazil signed the slave abolition treaty with Britain in 1826 as the price for recognition. Britain abolished slavery in her colonies in 1833 and, as the dominant naval power, enforced the ban on transport of slaves on the open seas (Adrian Desmond, James R. Moore, *Darwin's Sacred Cause: How a Hatred of Slavery shaped Darwin's views on Human Evolution*, 2009). Darwin wrote in *Descent of Man* (1871): "The main conclusion arrived at in this work, namely that man is descended from some lowly organized form, will, I regret to say, be highly distasteful to many. But there can hardly be a doubt that we are descended from barbarians." In his "Journal" (1845), he wrote:

> I thank God, I shall never again visit a slave-country (Brazil). To this day, if I hear a distant scream, it recalls with a painful vividness my feelings, when passing a house near Pernambuco, I heard the most pitiable moans, and could not but suspect that some poor slave was being tortured, yet knew that I was as powerless as a child even to remonstrate...Near Rio de

Janeiro I lived opposite to an old lady, who kept screws to crush the fingers of her female slaves...And these deeds are done and palliated by men, who profess to love their neighbours as themselves, who believe in God, and pray that his Will be done on earth! It makes one's blood boil, yet heart tremble, to think that we Englishmen and our American descendents, with their boastful cry of liberty, have been and are so guilty...Against such facts how weak are the arguments of those who maintain that slavery is a tolerable evil.

William Carpenter in 1844 asked: "Will anyone affirm that there is more difference between a Negro and a Caucasian, than between a greyhound and a mastiff; or that the education which, continued through a succession of generations, develops certain faculties and habits in the dog, shall be less effectual in man?" Ettiene Serres (1786–1868), a French embryologist, argued that the developing brain passed through a "woolly haired" phase, then one typical of an "intermediate Malay and American person, after that a Mongolian, and finally maturing as a Caucasian type" (after "ontogeny follows phylogeny" of Haeckel, Chapter C5).

Francis Galton (1822–1911), grandson of Erasmus Darwin, wrote *Hereditary Genius* in 1869, *English Men of Science: Their Nature and Nurture* (1874), *The History of Twins* (1875), and *Inquiries in Human Faculty and its Development* (1883). He summarized that there was no evidence that characters are transmitted in transfused blood; he coined the term "regression toward the mean" and began a classification of finger prints. He argued that intelligence is a function of skull size and shape and, more important, is heritable.

When does an interest in other peoples reflect a laudable curiosity and when is it demeaning and racist? Saartjie Baartman, advertized as the "Hottentot Venus," was displayed in England in the 1830s; for a fee, women could view her enlarged buttocks and thighs. Staetopygia, the exaggerated deposition, especially in women, of subcutaneous fat on the buttocks and thighs is characteristic of the Khoisan and Bantu. It is captured in the paleolithic sculpture, Venus of Willendorf, far from Southern Africa.

Joice Heth, an American slave, was exhibited by P.T. Barnum in 1836. The Hamburg Tierpark (zoo) sponsored an "anthropological-zoological" display of Lapps acting out "daily life" with reindeer in 1874. The Congolese pygmy, Ota Benga, was displayed in the Bronx Zoo in 1906. Today, throughout the world there are numerous exhibits of native crafts. On the Blue Ridge Parkway in Virginia, folks in reconstructed log cabins and period attire reenact the crafts of early mountain settlers in the "Pioneer Village"; they should not use their iPads during reenactments.

Even philosophers weighed in. David Hume said:

> I am apt to suspect the Negroes to be naturally inferior to the Whites. There scarcely ever was a civilized nation of that complexion, nor even any individual, eminent either in action or in speculation. No ingenious manufacture among them, no arts, no sciences.

Immanuel Kant contributed: "The yellow Indians do have a meager talent. The Negroes are far below them, and at the lowest point are a part of the American people." Not to be outdone, Georg Wilhelm Friedrich Hegel wrote: "Africa is no historical part of the world." Blacks have "...no sense of personality; their spirit sleeps, remains sunk in itself, makes no advance, and thus parallels the compact, undifferentiated mass of the African continent." — good examples of critical analysis.

Cesare Lombroso expressed the commonly held view that physical characteristics identify inborn criminals, *The Criminal Man* (1876). Alexander Winchell in *Sketches of Creation* (1878) suggested that Adam descended from earlier humans who might have been black. Virchow in 1885 argued that Europeans are a "mixture of various races" and he "...struggle(d) against any theory concerning the superiority of this or that European race."

An editorial in *Scientific American*, May 1911, titled *Science of the Unfit: The Science of Breeding Better Men* described Ada Juke as the "mother of criminals."

From her were directly descended one thousand two hundred persons. Of these, one thousand were criminals, paupers, inebriates, insane, or on the streets. That heritage of crime, disease, inefficiency and immorality cost the State of New York about a million and a quarter dollars for maintenance directly.

Karl Pearson lamented:

...in Great Britain 25 per cent of the population (and that the undesirable element in England) is producing 50 per cent of English children...although we may not be able directly to improve the human race as we improve the breed of guinea pigs, rabbits or cows...yet the time has come when the lawmaker should join hands with the scientist, and at least check the propagation of the unfit...The Eugenics laboratory founded by Sir Francis Galton and the American Breeders' Association have done much to clear away the popular prejudices inevitably encountered in such educational work and to prepare the ground for legislative action.

Teddy Roosevelt about 1890 was quite circumspect: "I don't go so far as to think that the only good Indians are dead Indians, but I believe nine out of them are and I shouldn't like to inquire too closely into the case of the tenth." Rudyard Kipling in 1899 coined the phrase "The White Man's Burden." Otto Weininger in 1906 stated: "A genius has perhaps scarcely ever appeared amongst the negroes, and the standard of their morality is almost universally so low that it is beginning to be acknowledged in America that their emancipation was an act of imprudence." Hannah Arendt in *The Origins of Totalitarianism* (1951) described the genocides in Namibia (1904–1907) and in Armenia (1915–1917) as part of the process of legitimizing imperialist conquests.

Paul Ehrlich in 1964 described two clines distributed discordantly — melanin north and south — and, in contrast, beta-S hemoglobin which radiates outward from specific points (Chapter C11). Neil Reisch (2002) said: "Both for genetic and non-genetic reasons, we believe that racial and ethnic groups should not be assumed to be equivalent, either in terms of disease risk or drug response..." "Whether African Americans, Hispanics, Native Americans, Pacific Islanders or Asians respond equally to a particular drug is an empirical question that can only be addressed by studying these groups individually."

Pearce and Dunbar in "Latitudinal variation in light level drives human visual system size" (2011) measured 73 200-year-old skulls from 12 different populations. The volume of the eye ball sockets ranged from 22.5 to 27.0 cm^3 (1/3 of the socket is eye ball); they found that eye socket orbits are larger at higher latitudes. Noback, Harvati and Spoor in "Climate related variation of the human nasal cavity" (2011) found the nasal cavities larger at higher latitudes and suggested a selective advantage to warming and humidifying cold, dry air. Are these studies of phrenology racist?

Herrnstein and Murray in the *Bell Curve* (1994) wrote: "The debate about whether and how much genes and environment have to do with ethnic differences remains unresolved." They concluded that:

Intelligence is largely (40% to 80%) genetically heritable and that no one has so far been able to manipulate IQ long term to any significant degree through changes in environmental factors — except for child adoption — and in light of their failure such approaches are becoming less promising. Epping *et al.* in "Parasite prevalence and the worldwide distribution of cognitive ability" (2010) reviewed the literature and found that national IQs correlate positively with: enrollment in secondary school, gross national product, and temperate climate and negatively with illiteracy, agricultural labor, and inbreeding. They argued that childhood infections cause lower oxygen tensions and injure the developing brain.

Jim Watson in an interview with The Sunday Times, 15 October 2009, stated that he was "...inherently gloomy about the prospect of Africa because "...all our social policies are based on the fact that their intelligence is the same as ours; whereas, all the testing says not really." He said there was a natural desire that all human beings should be equal but "...people who have to deal with black employees find this not true." He wrote in *Avoid Boring People: Lessons from a Life in Science* (2009):

> There is no firm reason to anticipate that the intellectual capacities of peoples geographically separated in their evolution should prove to have evolved identically. Our wanting to reserve equal powers of reason as some universal heritage of humanity will not be enough to make it so.

There are genetic differences among geographically distinct groups within the same species, including *H. sapiens*. In what ways, if any, should these differences inform political and educational policies?

C16

Information

Maxwell's Demon

high   entropy   low

How much information is available to the Demon?

Boltzman's definition of entropy, $S = k \cdot \ln W$, as chiseled on his tombstone, and the subsequent development of statistical thermodynamics was one of the crowning achievements of 19th century physics. During the first half of the 20th century communication in general and radio in particular enjoyed incredible advances. During the First and especially the Second World Wars radio was the focus of code making and breaking. Claude Shannon, in 1948, published *A Mathematical Theory of Communication*, one of the most influential papers in the history of science. Information, by his fundamental definition, can be equated with entropy. Further, he emphasized the difference between information and meaning, which is dependent on context and on the sensibility of the recipient. The sequences 417 and 911 have the same information content but for most recipients, at least in the United States, 911 has much more meaning.

Ever greater amounts of data are available from DNA sequencing. One searches for protein encoding genes, regulatory elements, and correlations with various phenotypes. The entire concept of a gene must be revisited. One of the fundamental questions is how to analyze this data and extract meaning. Are these traditional definitions of entropy and of information relevant to contemporary biology?

As summarized in Chapter B9, the formulation of the first law of thermodynamics, conservation of energy, depended on characterizing ΔU, change in all forms of energy — kinetic, chemical, electrical, potential, and thermal — as state functions and relating them to ΔW ("external mechanical" work done on the system) $+\Delta Q$ (heat flow). Herman Helmholtz (1821–1894) formulated the conservation of energy in 1847 and credited Mayer and Joule for their earlier statements, e.g.: "Energy can be neither created nor destroyed." — Mayer, 1841.

Sadi Carnot (1796–1832) in *Reflections on the Motive Power of Fire and on Machines Fitted to Develop that Power* (1824) gave these definitions: A heat engine is a device that takes heat from a hot source and performs mechanical work. A refrigerator is a device that transfers heat from a cold body to a hotter body. These and several other related statements resulted in the second law of thermodynamics: *It is impossible to construct a device that, operating in a cycle, will produce no effect other than the transfer of heat from a cooler to a hotter body.* Or, *it is impossible to construct an engine that, operating in a cycle, will produce no effect other than extraction of heat from a reservoir and the performance of an equivalent amount of work.* Students have their own versions, among those printable: *Things just go from bad to worse.*

James Clerk Maxwell (1831–1879) and Ludwig Boltzmann (1844–1906) developed the theory that describes the distribution of velocities of molecules, or atoms, of gas in a fixed volume under defined temperature and pressure. This distribution is the sum of micro-states of the system. Boltzmann established statistical thermodynamics with his formulation: $S = k \cdot \ln W$; k (the Boltzmann constant) $= 1.38 \cdot 10^{-23} \text{J} \cdot \text{K}^{-1}$. W is the Wahrscheinlichkeit (probability); $W = N! / \Pi_i N_i!$ N is the number of possible microstates and $N!$ is the frequency of occurrence of that microstate.

Charles Babbage (1791–1871) was the 11th Lucasian Professor of Mathematics at Cambridge University (1828–1839). He designed a *Differential Engine* to compute finite differences. It had 25,000 parts and weighed 13,600 kg. It could be programmed using punched cards as used in looms of the time. George Boole (1815–1864) wrote *The Mathematical Analysis of Logic* in 1847, establishing the field of symbolic logic; that is, logical propositions could be expressed by means of algebraic equations.

Vannevar Bush (1890–1974) was a professor in the Department of Electrical Engineering at the Massachusetts Institute of Technology from 1919–1932. In 1927 he constructed an analog computer, the *Differential Analyser*, to solve differential equations with up to 18 variables. He became president of the Carnegie Institution of Washington in 1939 and provost of MIT in 1949. He directed the National Defense Research Council, which became the Office of Scientific Research and Development in 1941 and led the Manhattan Project until 1943. In his report, *Science, The Endless Frontier*, to President Truman in July 1945, he described basic research as "the pacemaker of technological progress" and recommended the creation of what would become the National Science Foundation in 1950. Francis Bacon (Chapter A6) would have been gratified to see his "works" implemented.

Norbert Wiener (1894–1964) received his Ph.D. from Harvard in mathematical logic. He then studied with Bertrand Russell and G. H. Hardy at Cambridge and David Hilbert and Edmund Landau at Göttingen. He became professor at MIT, where he investigated information theory independently of Claude Shannon. He won the National Book Award for *God and Golem* in 1965.

John von Neumann (1903–1957) received his Ph.D. in mathematics in 1925 from the Pázmány Péter University in Budapest. He worked at the Institute of Advanced Study at Princeton with Albert Einstein and Kurt Gödel. He wrote *Mathematische Grundlagen der Quantenmechanik* (*The Mathematical Foundations of Quantum Mechanics*) in 1932. His *Theory of Games and Economic Behavior*

(1944) was lauded by Paul Samuelson: "Neumann was the only mathematician ever to make a significant contribution to economic theory." He consulted on the EDVAC project and the construction of the first electronic computer, the ENIAC, at the Moore School of Electrical Engineering at the University of Pennsylvania. He wrote a self-reproducing computer program, the first computer virus, in 1949.

Claude Elwood Shannon (1916–2001) worked with Vannevar Bush on his differential analyzer. He was one of the founders of digital circuit design and demonstrated that electrical application of Boolean algebra could resolve any logical, numerical relationship, "A Symbolic Analysis of Relay and Switching Circuits," *Trans. Am. Inst. Elec. Eng.* (1938). As a fellow at the Institute for Advanced Study, he met von Neumann in 1940. He went to Bell Laboratories to work on fire control systems and cryptography. There he met Alan Turing during his visit to Bell labs in 1943. Shannon's *A Mathematical Theory of Communication* in *Bell System Technical Journal* (two parts, July and Oct, 1948) and *Communication Theory of Secrecy Systems* in *BSTJ* (1949) essentially defined information theory. Shannon and Weaver in the 1951 *Prediction and Entropy of Printed English* in *The Mathematical Theory of Communication* defined information as $S = -_{i=1}\sum^n P_i \cdot \ln_2 P_i$ in which n = number of states; P_i probability of the ith state occurring. For example:

n	P_i	$\ln_2 P_i$	S	Example
2	$\frac{1}{2}, \frac{1}{2}$	$-1, -1$	1 bit	one toss of a fair coin
4	$\frac{1}{2},\frac{1}{2},\frac{1}{2},\frac{1}{2}$,	$-1, -1, -1, -1$	2 bits	two tosses of fair coin
6	$6 \times \frac{1}{2}'$s	$6 \times -1'$s	3	three tosses

n	P_i	$\ln_2 P_i$	S	Example
4	$4 \times \frac{1}{4}$'s	4×-2	2	ATCG, one nucleotide, if all 25%
12	$12 \times \frac{1}{4}$'s	12×-2	6	one triplet
6	$6 \times 1/6$'s	6×-2.6	2.6	one roll of a fair dice
12	$12 \times 1/6$'s	12×-2.6	5.2	two rolls
27	27×27's	27×-4.7	4.7	one character of our alphabet
54	54×27's	54×-4.7	9.4	two (26 letters and 1 space)

Frequency of usage in English, P_i

Letter	P_i	$P_i \ln_2$
a	0.0805	−3.67
b	0.0162	−6.01
c	0.0320	−5.02
x	0.0020	−9.06
y	0.0188	−5.82
z	0.0009	−10.22

The demonstration of the equivalence of entropy and information is foundational for communication theory and practice; it was a major achievement of 20th century science. There is a fundamental distinction between information and meaning; the latter depends on context.

As noted in Chapter B9, there are nominally predictable systems in which the slightest disturbance of initial conditions makes the system appear chaotic over a long time, the so-called butterfly effect. This is described as the Kolmogorov–Sinai (K–S) entropy. Systems whose K–S entropy is zero can be predicted exactly. Those with a non-zero K–S entropy are not wholly predictable; the tiniest perturbation at the level of atomic vibration can quickly perturb the system. They include chaotic systems.

Is information theory relevant to biology? Does this perspective clarify basic questions such as: Is all of the information in an egg contained within its DNA? How does the information in an egg determine the adult organism? What is the relevance of information to genomics, proteomics, connectomics, metabolomics? What in biology gives meaning to information?

C17

Origin of Life

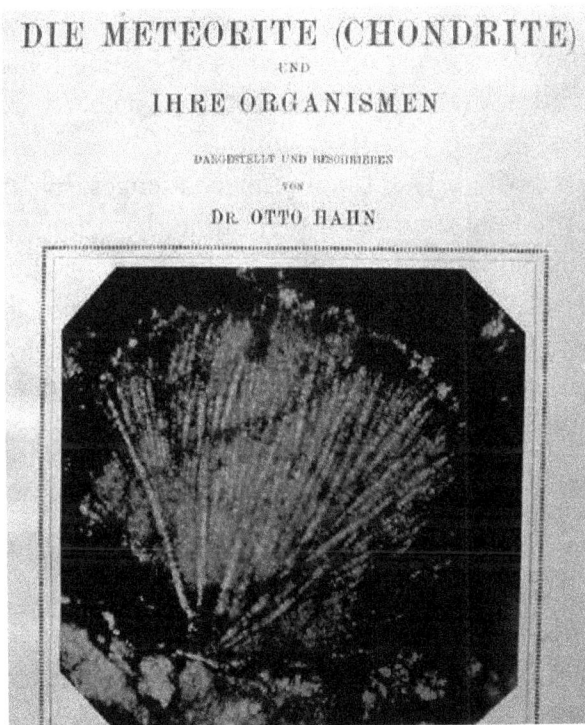

Darwin carefully avoided discussing the origin of life, just mentioning a "...warm little pond..." in a letter to Joseph Hooker in 1871. Origin(s) is now considered a legitimate field of research. Astronomers have identified over a hundred planets that are inferred to have (had) water and temperatures in the range of 0 to 100°C. These planets, judging from the precedents of our own

Earth, have a very heterogeneous geology and hydrology, that is, a wide range of conditions for chemical reactions. Over a hundred small, organic molecules have been identified in interstellar space. These organics can be accumulated on dust and delivered to planets in adequate quantities to support further chemical reactions. There are many mineral surfaces that concentrate organics from dilute solutions and might catalyze further reactions. The energy of their "sunlight" can vary over their "day." Local systems will not just come to steady state, or near equilibrium. These conditions can cycle or pump energy through the system, thereby creating negentropy, or order, on a microscopic scale. One can safely assume that all of these conditions obtained on earth and on many similar planets.

There are two, at least, outstanding challenges that are now subject to experimentation, not to determine the origin of life but at least to establish plausibility and focus for future observations and simulations. All cells of either unicellular or multi-cellular organisms have an inside, whose composition is different from the outside, a concept that can be traced back to the *mileau interieur* of Claude Bernard (Chapter C8). The chemical reactions inside these cells are subject to regulation. This distinction requires the formation of something resembling a membrane or structured nanopatches of surface minerals that have differing affinities for various "substrates." There must be some sort of gradient across which energy and matter can flow.

The realization that some oligomers of ribonucleic acid (RNA) can both encode and transmit information and serve as catalysts gave rise to reasonable models of an "RNA world" that preceded our DNA world. However, RNA is difficult to synthesize directly from the hundred small organics that might be available on a candidate planet. It is reasonable to search for simpler polymers or surfaces that preceded RNA but served the dual functions of catalysis and information transmission.

One is an odd number. If life originated via these chemical processes, it seems very strange that it should have occurred only once, as opposed to a few million times. If only once, is that real proof of God? Life may have originated, *de novo*, on many planets, or may have been transferred planet to planet, as fragments are sometimes "splashed" off from bolide collisions, i.e. "panspermia." If there are multiple origins of life, is our physiology, sociology, and intelligence unique? If so, is *that* real proof of God?

Before addressing these questions, one asks, what is life? A living system should capture energy from the environment and use it to create order, or negentropy. A living system should be able to replicate itself with reasonable fidelity. However, perfect fidelity would not permit evolution.

Water is unique. Each molecule of H_2O can donate and receive two hydrogen bonds. It can dissolve a wide range of ions and molecules. It can engage in a wide range of chemical reactions but is not too reactive or corrosive. Its maximum density occurs at 4°C; ice floats. Few can imagine life without water — hardly a defensible argument from a philosophical perspective.

Carbon is unique. It can be tetra-valent with sp^3 orbitals or tri-valent with sp^2 orbitals. Nearly 10,000,000 organic compounds with an enormous range of properties have been discovered or synthesized. Few can imagine life without carbon.

The surfaces of crystals are usually regular in two dimensions; this lattice is characterized by the lengths, **a** and **b**, and by the inter-axial angle, γ. Depending on the mineral composition, an enormous variety of reactive groups at various orientations can be displayed on these surfaces. Discontinuities or grain boundaries on these surfaces provide a range of catalytic niches and characteristics. A very broad range of reactions might be catalyzed. Energy, either from geothermal or solar sources, might periodically flow past these surfaces and be used to generate order.

Only within the past decade have observations of other stars become so sensitive as to detect the transits of their orbiting planets and to infer that some have temperatures from 0 to 100°C. One can safely infer the existence of many planets with temperatures, compositions, and histories, similar to ours. Most planets rotate about their own axes as they orbit their suns. Many vents or geysers release energy and/or material cyclically, e.g. "Old Faithful" in Yellowstone. It is plausible to imagine systems through which energy flowed. Each cycle might explore a range of conditions and permit the accumulations and disposition of rare products.

Astronomers have also detected spectral lines that can be assigned to over a hundred small molecules or ions present in interstellar space. These can accumulate on dust particles. One can safely assume the existence of planets that were, are, or will be similar to Earth in general composition and temperature.

Although the actual mechanisms have yet to be identified, it seems quite reasonable to assume that a reasonable range of compounds, organic and otherwise, were available in reasonable concentrations in Darwin's "warm little pond." Energy — either light, heat gradient, or geologically formed compounds, e.g. sulfides — was readily available.

The "RNA world" hypothesis (Gilbert, 1986) noted that ribozymes can catalyze reactions and that genetic information is encoded in both single-stranded and double-stranded RNA viruses. RNA is inferred to have preceded DNA. Single polymers of RNA both catalyze reactions and store genetic information. Still, it is an enormous leap from a few hundred small organic molecules and energy cycles to ribozymes. The most simple linear code would involve two (chiral) symbols, or molecules; e.g. → BBABAABABBAAAB → would carry information. In a simple scheme it would be copied, with occasional errors, to → BBABAABABBAAAB →. Earth's genetic code differs in three ways. It uses four, not two, symbols or molecules — adenine thymine guanine, and cytosine. Second, these symbols can be modified, e.g. methylated to modify their access to "readers." Third, the four letter message, e.g. → CCTCAATGGGTCAT → is not directly copied to → CCTCAATGGGTCAT → but instead to its complement ← GGAGTTACCCAGTA ←. The complement is then copied back to its complement, the original → CCTCAATGGGTCAT →. The transition from → BBAB, . . . → CCTC . . . might be modeled; however, getting from a few hundred or a few thousand small organic molecules to → BBAB... is no mean challenge for that warm little pond.

Aristotle (384–322 B.C.) knew that mice emerged from dirty hay and flies from rotting food and aphids from drops of dew. Spontaneous generation seemed quite reasonable. Francesco Redi in 1668 demonstrated that maggots do not appear on meat if flies are prevented from laying their eggs. Louis Pasteur in 1864 compared growth in two boiled broths — one exposed to air and the other not (Chapter C6) — and concluded: "Never will the doctrine of spontaneous generation recover from the mortal blow struck by this simple experiment."

But what next? Organic compounds, including, oligomers of amino acids or carbohydrates might be generated and segregated. But how could their presence "control" or influence the generation and/or degradation of other molecules somehow related to themselves?

Alexander Oparin, in *The Origin of Life* (1924), argued that atmospheric oxygen prevents the synthesis of some purines and amino acids. That "origin" needed a reducing atmosphere — no problem. Our early atmosphere contained little oxygen; reducing niches still exist. In the Miller–Urey experiment (1952), a mixture of hydrogen, water, methane, and ammonia was heated, refluxed, and subjected to frequent electrical sparks. After several days, over 10% of the carbon was in organic compounds, including amino acids. Sidney Fox dried solutions of amino acids. They formed cross-linked polymers that in turn aggregated into "proteinoid" globules. Joan Oró synthesized adenine from aqueous solutions of ammonium cyanide solutions. Kahr *et al.* (2007) reported their experiments that tested the idea that crystals can act as a source of transferable information. "Mother" crystals of potassium hydrogen phthalate with imperfections were cleaved and used as seeds to grow "daughter" crystals from solution.

> But if (and oh what a big if) we could conceive in some warm
> little pond with all sorts of ammonia and phosphoric salts,
> light, heat, electricity *et cetera* present, that a protein compound
> was chemically formed, ready to undergo still more complex
> changes . . .
> —Darwin was on the right track; letter to Joseph Hooker
> (1871).

At a minimum, bio-organic chemistry would require: carbon (C), oxygen (O), hydrogen (H), and very probably nitrogen (N). Phosphorus (P) and sulfur (S) are found in all biochemical systems and add tremendous flexibility. Most "ocean" environments contain sodium (Na^+), potassium (K^+), magnesium (Mg^{2+}), and calcium (Ca^{2+}) cations and, for balance, chloride (Cl^-) anions. Cells would soon have to learn to deal with them; then probably become dependent on them. At least some organisms, but certainly not all, require bromine (Br), iodine (I), silicon (Si often omitted from such a list) and the 11 metals: vanadium (V), chromium (Cr), molybdenum (Mo), tungsten (W), manganese (Mn), iron (Fe), cobalt (Co), nickel (Ni), copper (Cu), zinc (Zn), cadium (Cd). No models for the initial ur-chemistry include elements beyond C, O, H, and N.

Researchers at the J. Craig Venter Institute identified 475 protein encoding genes in the 580 kilo-base genome *Mycoplasm genitalium*. They have created the minimal genome organism, *M. laboratorium*, that has 382 protein encoding genes and 43 RNA structural genes; it can grow and divide in a controlled environment. Development of these techniques will permit the creation of designer bacteria. Whether such minimal bacteria reflect a stage in the evolution of contemporary bacteria remains problematic. Just to complicate matters, Nadège *et al.* recently sequenced the genome of a pandoravirus, 2,500,000 bp DNA, encoding >1,500 genes; why its host *acanthamoeba* tolerates it remains unknown.

Carl Friedrich Gauss in 1820 supposedly suggested that a giant right triangle and three attached squares could be drawn on the Siberian tundra. The outlines of the shapes would have been 15 km-wide strips of pine forest; the interiors could be rye or wheat —

quite visible to a telescope on Mars. Carl Sagan in his novel *Contact* explored in some depth how a message might be constructed to allow communication with an alien civilization, starting with prime numbers. Others have suggested sending the value of π or of e, base of the natural logarithms, encoded in binary, essentially dots and dashes. Pioneer 10 and Pioneer 11, launched in 1972 and 1973, carried plaques depicting the location of the Earth in its solar system and galaxy as well as the human body. Voyager (1977) carried two golden records inscribed with pictures and sounds from Earth. The Search for Extraterrestrial Intelligence Institute monitors several radio wave lengths to identify non-random signals from other solar systems. They have completed sophisticated evaluations of how messages might be recognized as well meaningful messages to be sent. Where is everyone?

One is an odd number. Surely life of some form exists on other planets. Surely some of that life evolved to grind lenses and to generate radio signals. Or is it so sure?

Section D

Society and Science: Overview

Most aspects of contemporary society have components involving science, often biology, as broadly characterized in these chapters. Practicing scientists participate in discussions of policy, politics, and philosophy; however, most of those involved in setting policy have little formal training in science. How can someone, perhaps a scientist himself, not familiar with the science under discussion make use of advice and information in questions of policy? As important is the question of how allocation of resources and the intellectual climate of society affect science.

The following chapters are hardly comprehensive; a few topics have been chosen to illustrate the broad range of interactions between society and science. Implicit in these examples is reciprocity, not dominance.

D1. Integrity —
   addresses the development of standards of attribution, priority, and copyright.
D2. Informed Consent —
   addresses the development of ethics for use of people in experimentation.
D3. Faith —
   explores the role of religion as exemplified by the current debate over intelligent design.

D4. Art —

   notes many examples of science having inspired art and asks how art inspires science.

D5. Global Warming —

   considers a contemporary controversy and asks whether it can be understood in these terms.

D6. Free Will —

   asks in what way human decisions are influenced by genes, by the environment, and by chance.

D1

Integrity

RISE OF THE RETRACTIONS

In the past decade, the number of retraction notices has shot up 10-fold (**top**), even as the literature has expanded by only 44%. It is likely that only about half of all retractions are for researcher misconduct (**middle**). Higher-impact journals have logged more retraction notices over the past decade, but much of the increase during 2006–10 came from lower-impact journals (**bottom**).

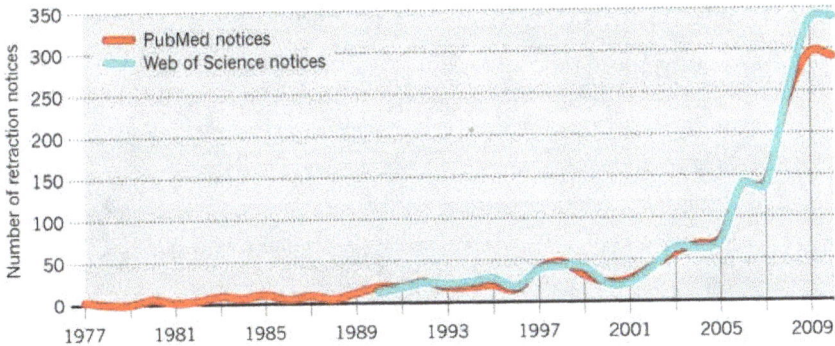

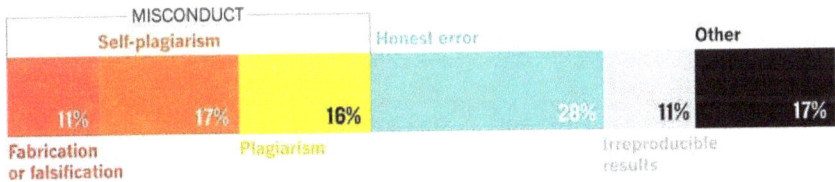

Integrity, or the lack thereof, can be expressed in many ways in pure science as well as in engineering and medicine. Many academicians feel that their norms and practices set a high standard for society at large. Does this stand up to scrutiny? This chapter touches on

287

several related topics, each of which might be the subject of a great tome or tomes, depending on one's perspective.

- What are, and were, standards regarding falsification, plagiarism, and attribution in academic science?
- How and why have these standards changed over time and place?
- To what extent are, or ought to be, these standards the same in the applied sciences?
- To what extent are, or ought to be, these standards the same in society at large, and how do they influence one another?

No contemporary society condones theft or breach of contract, yet the criminal and civil codes run to volumes. Common sense and consensus are not adequate for the administration of justice — "nation of laws, not men." Hunter gatherers have precedents and oral tradition. Until the 1800s, the "administration of science" was an oxymoron. There were indeed conflicts of priority and attribution. Newton assumed an inverse square law when formulating equations to describe the orbits of the planets about the sun (Chapter B4). Hooke claimed to have suggested this essential idea to Newton, who replied: "If I have seen further it is by standing on the shoulders of Giants." Some have suggested that this was hardly an acknowledgement but a snide reference to Hooke's kyphosis and resulting hunchback. Lavoisier claimed credit for the discovery of *principe oxygine* actually first discovered and characterized by Joseph Priestly in 1774 (Chapter B6). In 1782, Lavoisier claimed credit for the experiments of Cavendish in which he demonstrated the combination of two equivalents of hydrogen and one equivalent of dephlogisticated water, oxygen, to form water. These lapses were not the reason for his beheading in 1794. Nor is the guillotine the appropriate response today except, as argued by deniers, for those involved in climate research (Chapter D5). Lister described his *Antiseptic Principle of the Practice of Surgery* (1867). He referred to Pasteur's work (Chapter C6) and readily acknowledged: "Without Semmelweis (and Holmes), my achievements would be nothing."

These standards, without being specifically codified, seem to have been generally accepted by the time of the scientific revolution, the 1600s. The secrecy and intentional obscurity of some alchemists was one of the reasons that they were held in distrust by the general community (Chapter B5). One can cite specific transgressions, as above; however, as with any historical study, one must be cautious to extrapolate to a general pattern.

Certainly scientists had heated debates; reputations might be ruined or enhanced. However, there were few formal guidelines or procedures. It was generally accepted that secondary schools in general and universities in particular usually honored high standards and, especially important, judged their students and colleagues fairly.

It is difficult to document or to quantify the three major sins of research — falsification, plagiarism, and failed attribution. The list can be extended: failing to declare conflicts of interest, publishing the same, or similar, data more than once (exclusive of review articles), selective reporting, and honorary authorships. In response to "anonymity assured" surveys, depending on the wording and list of sins, 10% to 50% of scientists around the world admit to some transgressions.

Jonah Lehrer wrote three books (*Proust was a Neuroscientist*, 2007, Chapter D4) and over 20 articles for *The New Yorker*, *The Guardian*, *The Wall Street Journal*, and *Wired.com*. He was charged by several people with self-plagiarism, recycling content, and fabricating quotations. Houghton, Mifflin, Harcourt withdrew his *How We Decide* from market. Lehrer acknowledged his transgressions and apologized. Doris Kearns Goodwin, presidential historian and frequent commentator on the Public Broadcast Services Newshour, seems to have recovered from charges of plagiarism and self-plagiarism. One wonders how many journalists and scholars could pass this level of scrutiny? Perhaps more importantly, are such standards applied equally?

Retraction Watch, a blog authored by Ivan Oransky and Adam Marcus, surveys the millions of publications listed in PubMed and attempts to assign the reasons for retractions, often not supplied by the authors, nor demanded by the journal. In great simplification, retractions are more or less equally distributed, normalized to number of publications, worldwide. It is difficult to establish the real reason for the retraction; however, Oransky and Marcus estimate that half reflect some sin by one of the authors.

Some journal editors have been reluctant to participate in such evaluations or to demand specific statements of chastity from contributors. Some journals have been found to use "tricks" to enhance their rankings in citation indices.

Of greater concern, the normalized frequency of retractions and inferred fraud, as opposed to honest mistakes, has increased markedly during the past decade. Although not documented, nor inferred in these studies, very probably the incidence is much higher in biology than in physics. One interpretation is that there are often many more variables and the data are less precise in biology; it is more difficult for another investigator to repeat exactly the experiment in question. Other interpretations are less charitable.

The vast majority of academic research in the United States is funded by the federal government. Quite reasonably those results should be published in a timely manner and basic data archived and available. Critical reagents, strains of bacteria, and tissue cultures should be available to other labs for nominal handling fees. The results of this research should benefit the scientific community as well as its benefactors, the taxpayers.

As funding for universities tightens, researchers are under mounting pressure to conduct "translational research" that results in profits and patents for the host institution. Researchers are encouraged to establish for-profit "start-up" companies that depend on the scientists' expertise and discoveries made from the federally supported research. It is a challenge both to researchers and to administrators of good faith to strike an appropriate balance.

Some argue that this for-profit attitude within academe places subtle, or not so subtle, pressures on faculty, especially in engineering and medicine, to be profitable by getting patentable results and/or by the "indirect costs" that are paid to the host institution by the granting agency. This might compromise the commitment to basic science and fundamental discoveries that were not, and could not have been, anticipated or included in the "significance" section of that research proposal.

The Department of Health and Human Services had a fiscal year 2013 budget of $982 billion. The Public Health Services (National Institutes of Health), through intra- and extramural programs, spends about $31 billion per year for health related research and development. Its Office of Research Integrity (ORI) carries out its responsibility by: "Developing policies, procedures and regulations related to the detection, investigation, and prevention of research misconduct and the responsible conduct of research" as elaborated in a long list of more specific charges listed on the ORI home page (381:1097–98, 2013).

The ORI has mandated a course in research ethics to be taken by graduate students and post-doctoral fellows in institutions receiving PHS funding. Its format and detailed content is, wisely, not specified; however, the course should address:

a) conflict of interest — personal, professional, and financial
b) policies regarding human subjects, live vertebrate animal subjects in research, and safe laboratory practices
c) mentor/mentee responsibilities and relationships
d) collaborative research including collaborations with industry
e) peer review
f) data acquisition and laboratory tools; management, sharing and ownership
g) research misconduct and policies for handling misconduct
h) responsible authorship and publication
i) the scientist as a responsible member of society, contemporary ethical issues in biomedical research, and the environmental and societal impacts of scientific research.

There are relatively few documents that give explicit answers or guidelines to these topics and implied questions. The national and international science communities have a strong sense of appropriate behaviors. The course, with discussion groups, should distill and clarify the collective wisdom of the community. It is not easy to codify or to express this wisdom but no less valuable to explore it. Many of these discussions are most meaningful when seen in context — "situation ethics."

Andrew Wakefield published a study in *Lancet* in 1998 that linked the combined measles, mumps, and rubella (MMR) triple vaccine to increased incidents of autism. He was found to have several conflicts of interest and to have manipulated the data. *Lancet* fully retracted the paper in 2010. Wakefield was found guilty of serious professional misconduct in 2010 by the General Medical Council of Britain and his name was struck off the Medical Register. Would a course in ethics have prevented his misconduct?

If a couple of dudes, let's call them Orville and Wilbur or Sergey and Larry, are working in a garage start-up, their communications are verbal and their documentation is scanty; all sealed with a handshake. Is it in society's best interest to codify these interactions? In contrast, well-established companies like DuPont or Ford Motors have elaborate protocols for keeping records. These are essential for patent applications, or circumventions. Employees sign non-disclosure contracts, similar to security clearances in government. Some of this work is published, after applications for patents and evaluation of value to competitors.

The value of such "works" (Chapter A6) to society is well recognized and drove the development, an ongoing process, of copyright and patent law. Patents were issued in Italy and France in the 1400s; the first American patent was issued in Massachusetts in 1641 and by the U.S. government in 1790. The unanimous ruling by the U.S. Supreme Court in *Association for Molecular Pathology vs. Myriad Genetics* in 2013 that naturally occurring genes cannot be patented is of special interest to biologists. Many best practices in industry are driven by concern for patents. Most universities have staff well-versed in patent law. There are many fruitful collaborations between industry and academe. It requires a sophisticated infrastructure to monitor and administer so many moving parts.

Most industrialized nations have converged to similar standards within academe. In developing nations, practices of academic publication are nominally similar; however, there are relatively more examples of multiple publications of the same experiments. The standards — patent law, safety protocols, eminent domain, espionage — for engineering and for medicine may reflect local tradition and resources in developing nations. Both national companies and those with home offices in the West often pay bribes and fail to honor even minimal standards of worker safety and waste disposal. However, their practices are also converging toward best practices in the West, sometimes under treaty commitment and/or threat of sanctions.

The traditional relationships between student and teacher — no cheating and fair evaluations — are the same for scientists and for humanists. However, practices differ significantly for the post-baccalaureate level training of scientists and, to some extent, engineers and physicians. For over a century in the United States a doctoral degree, the Ph.D., has been required for science faculty in major universities and colleges. In contrast, only since about 1950 have the faculty of most schools of engineering had a Ph.D., as opposed to a Master's, M.S. or M.A. The norms and nomenclature in other countries vary but most now use the U.S. Ph.D. as a reference or equivalent.

Often humanists go directly from their doctorates to faculty positions or equivalents outside of the academy. Nearly all aspiring scientists spend an additional couple of years in a post-doctoral position, often supported by a federal grant to the mentor. Some fellowships are awarded directly to post-docs. The tenure of a post-doc has steadily increased over the past 50 years to an average of about three years, often extended to "jobs" as staff scientists or research assistants. This extension reflects either the granting of too many Ph.D.s, with a longer queue competing for faculty positions, or more training required to reach independence, or exploitation of cheap labor by the mentors. Interpretations vary depending on one's perspective as administrator, mentor, or mentee. A similar system of funding and mentoring obtains for graduate students. Ever more undergraduates are involved in research as they are better prepared and as mentors appreciate their assistance. This shift in training and demography puts increasing importance on the ethics of mentor–mentee relationships.

The average number of authors on publications has increased from usually one in 1910 to a median in biology of about four in 2010. Most of these publications bear the name of the lab director(s), usually listed last, and the student(s) (post-doc to undergraduate) who did the hands-on work, listed first or in descending order of contribution. A research, as opposed to a review, paper with a single author is now extremely rare. Frequently the authors hail from different institutions. This reflects the collaborative nature of much contemporary research. The participating labs have specialized resources or expertise and easy communication. Long gone are the heroic days when a graduate student worked alone on his (seldom her) "suggested" research project. Then upon receipt of his Ph.D., he assumed a faculty position, teaching undergraduates, and in his spare time continued his own research.

This shift in style has placed much greater emphasis on administrative talents, both in obtaining funding and in administering complex collaborations. Contemporary scientists require the skills of administrators and are often hired by industry for those talents. Science might once have provided haven for the eccentric, reclusive scientist, e.g. Cavendish (Chapter B6), but no longer. The training of doctoral candidates in science rightly emphasizes research. Whether formal instruction in administration should be part of the graduate curriculum is now seriously debated.

A greater fraction of publications in physics result from "big science," that is, consortia using expensive and scarce resources, e.g. synchrotrons and satellites. During the past two decades much of biology has become big science, e.g. genome sequencing projects and specialized facilities. There is enormous variation in these diversified systems; however, two generalizations merit comment.

The training of researchers remains basically a guild system in contrast to the instruction and evaluation of undergraduates or the employment of staff scientists. The apprentice is very dependent on the suggestions and the evaluations of the master. This can result in the formation of close bonds that last a lifetime or, on rare occasion, result in exploitation. Some practices are generally accepted, such as having a committee that reviews the candidate's progress annually. However, a great deal is left to the good judgment and goodwill of the mentor, e.g. authorship on publications from the entire lab or consortium and on "permission" for the post-doc to take aspects of graduate research on to her or his new lab. There is tendency to codify these procedures, as done by ORI for courses in research ethics. Whether, on average, these guidelines can replace or enhance general, ethical judgment remains to be seen.

Undergraduate and high school lab courses emphasize that one should record the observations, as made. Yet, courses are graded, and papers are published based on getting the "right" answer. Many, if not most, experiments "don't work"; the results should be recorded but not reported. This gives rise to the witticism "This is a typical result; it's the best one we got." To what extent should "wrong" results be discarded and at what level might this be considered dishonest? Further, in discarding wrong results, might one be overlooking unanticipated X-rays or penicillin? To what extent might one withhold results that are politically unwelcome and put at risk one's career?

On average scientists probably commit fewer crimes, or personal transgressions, than does the population at large, even the population of professionals. Does this relative civility reflect their professional standards and training or perhaps the fact that their results may be objectively tested? Does it serve them well if they enter the world of politics or of commerce?

Some scientists have strong political interests, e.g. conservation, that are related to their research. They must meet the challenge of presenting their analyses in a balanced manner. Their critics find it difficult to believe that they don't "spin" their results to support their politics.

It is hardly surprising that scientists are often very patriotic and serve their countries by developing defenses, medications, and weapons. Some of these countries are considered to have been evil, especially by their opponents. To what extent has a person who served his country compromised his integrity as a scientist?

Fritz Haber (1868–1934) was a chemist of Jewish background who received the Nobel Prize in Chemistry in 1918 for his development of a process for synthesizing ammonia, important for fertilizers and explosives. He also developed processes for deploying chlorine and other poisonous gases during World War I. The development and dropping of the atomic bomb by the United States is referred to as the loss of innocence for the physics community. The same might be said for dynamite and Alfred Nobel.

Scientists are admonished to assume responsibility for the societal impacts of their discoveries and inventions (including video games?). Easily said. A young assistant professor, working for tenure, identifies a gene that encodes a protein, whose mutant form is strongly associated with mental retardation. Beyond publishing her results, what is her responsibility, or that of her community?

D2

Consent

Mycobacterium leprae (leprosy) (Gerhard Armauer Hansen, 1895).

As discussed in the previous chapter, of all the issues of integrity concerning medicine, none is of greater importance, nor has broader implications, than informed consent. This chapter addresses several interrelated aspects.

What is "informed" and what is "consent"? Their definitions or characterizations have varied with circumstances.

The vaccination of James Phipps by William Jenner was accepted, if not considered heroic, in 1798. In the United States there were scores of instances of doctors "testing" patients, without their knowledge, and hardly their consent, with various inoculations and drugs. Under eugenics laws, over 7000 people were sterilized in Virginia (1924–1979). The infamous Tuskegee syphilis experiment resulted in the infection of, and/or the withholding of penicillin treatment from 600 poor Blacks in Alabama. Parallel syphilis experiments were conducted in Guatemala (1946–1948). The Nazis performed a range of experiments in their prison camps to evaluate war conditions and therapies; nearly all of the subjects died. The resultant Nuremberg code of 1948 was the first of several declarations codifying acceptable practices first in medical experimentation, then of various human rights.

Finally, one asks how the concept of informed consent can be extended beyond medical procedures, especially to the acquisition, analysis, and use of personal data.

Informed consent, like apple pie, can hardly be contested. However, both "informed" and "consent" are complex concepts whose meanings have changed over time and place.

Herophilos (335–~280 B.C.) performed the first recorded dissections of humans, executed criminals, as cited by Galen. There is no record of consent from the criminals or from their dearly beloved.

Edward Jenner (1749–1823) in 1796 vaccinated James Phipps, age eight, with cowpox pustules. He knew that Benjamin Jesty, a Dorset farmer, had so treated his wife and daughter to prevent smallpox (Chapter C6). Jenner had a reasonable expectation that no harm would come to Phipps and that he might well develop immunity to smallpox, which he did. Jenner was lauded as a hero. There is question as to whether Phipps was fully informed or whether he or his parents provided informed consent.

Louis Pasteur (1822–1895) had tested his newly developed rabies vaccine successfully on 11 dogs by 1885 before he treated Joseph Meister, age nine, who had been bitten by a rabid dog, a death sentence. The boy survived (Chapter C6). The actions of Jenner and of Pasteur, who was not a licensed physician, were not reviewed by an institutional ethics panel.

In 1880, Gerhard Hansen (1841–1912) tried, unsuccessfully, to infect a patient in Bergen, Norway, with *Mycobacterium leprae* as part of his study of Hansen's disease (leprosy). He lost his hospital position. Albert Neisser (1855–1916), in whose honor the "clap" bug was named *Neisseria gonorrhoeae*, infected patients (mainly prostitutes) with syphilis without their consent in 1900 in order to study the course of the disease. His actions were generally supported by the academic community. Public opinion was opposed; he was fined by the Royal Disciplinary Court in Germany.

These sorts of tests continued to recent years; a few examples illustrate the range of circumstances. Marion Sims in the 1840s performed surgery on slaves using no anesthesia; most of these women died from the resulting infections. To test his theory of trismus, restricted movement of the jaws, in infants he used a shoemaker's awl to move the skull bones of slave babies. Roberts Bartholow of the Good Samaritan Hospital in Cincinnati opened the skull of his cancer patient, Mary Rafferty, an Irish domestic, and inserted electrodes into her brain. In his research report he noted that she convulsed violently. Arthur Wentworth at the Children's Hospital, Boston, performed spinal taps on 29 children, without knowledge of their parents; he was curious whether the taps would be harmful. Henry Heiman in 1895 in New York infected two mentally disabled boys with *N. gonorrhoeae*.

Turning to adult males, in 1906, Richard Strong infected 24 Filipino prisoners with cholera unintentionally contaminated with *Yersinia pestis* (plague). All became ill; 13 died. From 1913–1951, Leo Stanley, physician at San Quentin Prison, transplanted testicles of executed prisoners or of animals into living prisoners.

Three doctors working at St. Vincent's House orphanage, Philadelphia, infected dozens with *Mycobacterium tuberculosis*; this resulted in permanent blindness for several. The kids were referred to as "material used." Hideyo Noguchi of the Rockefeller Institute injected 146 hospital patients with *Treponema pallidum* (syphilis); he was sued by parents of several child subjects.

In 1941, William Black inoculated a 12-month-old baby with *Herpes simplex* virus. Payton Rous, editor of the *J. Exp. Med.*, rejected the resulting manuscript as "... an abuse of power, an infringement of the rights of an individual, and not excusable because the illness which followed had implications for science." The MS was later published in *J. Pediatrics*. Francis and Jonas Salk, from the University of Michigan, sprayed *Orthomyxoviridae* (influenza) virus in the noses of patients in mental institutions. Rous wrote: "It may save you much trouble if you publish your paper ... elsewhere than in the *J. Exp. Med.* The journal is under constant scrutiny by the anti-vivisectionists who would not hesitate to play up the fact that you used for your tests human beings of a state institution. That the tests were wholly justified goes without saying."

From the 1940s–1970s, the Department of Medicine, University of Chicago, with U.S. Army and State Department collaboration tested the effects of malaria on prisoners of the Stateville Penitentiary. Alf Alving, U. Chicago, infected psychiatric patients at the Illinois State Hospital to test treatments from 1944–1946. In 1950 the U.S. Navy sprayed *Serratia marcescens*, thought harmless, over San Francisco resulting in many pneumonia-like illnesses. The family of one deceased sued; the federal judge ruled for the Navy; similar tests continued to 1969. In "Exercise Desert Rock" (1951), U.S. troops were intentionally exposed to radiation from tactical nuclear weapons to see the effects. The Central Intelligence Agency released *Bordetella pertussis* (whooping cough) over Tampa Bay; 12 died in the resulting epidemic.

Not to be outdone, the U.S. Army released mosquitoes infected with *Flaviviridae* (yellow fever) or with dengue fever over Savannah, GA, and Avon Park, FL, in 1956. Hundreds developed fevers, respiratory problems, and encephalitis. Army personnel pretended to be public health workers to photograph and perform medical tests on the victims, several of whom died. In 1966, the U.S. Army released harmless *Bacillus globigii* in the subway tunnels of New York to simulate a "Covert Attack with Biological Agents."

In 1950, Joseph Stokes, from the University of Pennsylvania, infected 200 female prisoners with *Hepatitis A* virus. From 1963–1966, Saul Krugman, at the Willowbrook State School, Staten Island, infected mentally disabled children with hepatitis; their parents were told of "vaccinations."

Chester Southam from the Sloan-Kettering Institute injected live cancer cells into healthy female prisoners of the Ohio State Prison in 1952. Ten years later he injected 22 elderly patients in the Jewish Chronic Disease Hospital in Brooklyn to "discover the secret of how healthy bodies fight the invasion of malignant cells." The New

York State medical licensing board placed him on probation for one year. American Cancer Society elected Dr. Southam vice-president in 1964.

Henrietta Lacks, age 31 in 1951, was diagnosed with an aggressive cervical cancer at the Johns Hopkins Hospital in Baltimore. Her cancer was biopsied, as was then standard practice, without her knowledge; she died six months later. Cells derived from her cancer were cultured. In contrast to all other human tissues up to that time, they grew and divided; their immortal progeny have been and still are used in thousands of labs around the world; their use has been referenced in 75,000 research papers. Recently the HeLa cell genome has been sequenced by labs in Hamburg and in Seattle. The reason for their immortality appears to be the insertion of the human papilloma virus genome on chromosome eight in the control region of an oncogene, permanently turning it on.

In 2010, Rebecca Skloot wrote *The Immortal Life of Henrietta Lacks*, after reviewing the basic science and interviewing several of Lacks' descendants. The latter were offended that they had had no voice in the disposition of their grandmother's cells and were concerned that the recently determined gene sequence might enable researchers and insurers to infer something about their own genomes without their permission. A civil agreement was negotiated by Francis Collins, director of the NIH. The family requested no compensation but instead reached an agreement that all published research involving HeLa cells, funded by NIH, should acknowledge Henrietta as the source of the cells. Further, a six-member review panel, with two members from her descendants, should review all requests for the HeLa DNA sequence and assure that these researchers commit not to disclose any information that might lend insight into the DNA sequence of her descendants.

Carrie Buck in 1910 was admitted as "defective" to the Virginia Colony for Epileptics and Feeble-minded (VCEF) in Lynchburg; her mother had been admitted to the VCEF in 1906. Carrie was classified as "feeble-minded" after birth of illegitimate child, also "feeble-minded," a product of rape by a relative of her foster family. In 1927, the U.S. Supreme Court in Buck vs. Bell approved the sterilization of Carrie Buck to prevent the birth of more "defective" individuals. Supreme Court Justice Oliver Holmes opined: "Three generations of imbeciles are enough." The Virginia Advisory Legislative Council in 1961 recommended "no change to the sterilization statute" citing "no substantial complaints."

The "Eugenical Sterilization Act", Virginia SB 281, was adopted in 1924 and repealed in 1979; similar laws were adopted in many other states. Over 7,500 people were sterilized (Largent, 2008; Dorr 2006). Half were "mentally ill," half "mentally deficient"; the distinction is not obvious, but just to be safe. The justification was maintenance of the "American race" and "traditional Southern identity." "Mountain sweeps, by local sheriffs rounded up "mongrels" and "worthless" whites.

The Public Health Service (PHS) in collaboration with the Tuskegee Institute studied the natural progression of untreated syphilis from 1932–1972. Six hundred black sharecroppers in Macon County, AL, were promised free medical care, meals, rides, burial insurance, and treatment for "bad blood." Three hundred and ninety-nine had syphilis; 201 did not. Taliaferro Clark drafted the initial plan, a six to nine month study; he retired in 1933. The study assumed a life of its own. Nurse Eunice Rivers, from Tuskegee, served as a link to the community for 40 years. Two hundred and fifty of the participants registered for draft in 1942; those with syphilis were diagnosed and were successfully treated. After the war, penicillin was generally available; however, "So far, we are keeping the known positive patients from getting treatment."

Peter Buxtun, a PHS venereal disease investigator, wrote a letter to the director of the Division of Venereal Diseases in 1966 expressing his concerns about the ethics and morality of the study.

The Centers for Disease Control supported by National (Black) Medical Association and American Medical Association urged that the study be continued until all of the subjects had died and been autopsied. In 1968, William Jenkins of the PHS founded a newsletter, *The Drum*, in which he called for an end to the study; he got no response. In 1972, Buxtun contacted the *Washington Star* and the *New York Times* who published the story on July 25 and 26, 1972. Following congressional hearings, a CDC and PHS *ad hoc* advisory panel recommended termination. In 1972, 74 of the 400 subjects were still alive; 28 had died of syphilis; 100 had died of complications; 40 wives were infected; 19 children were born with congenital syphilis. The follow up 1979 Belmont Report led to the establishment of Office for Human Research Protections and passage of federal laws, requiring Institutional Review Boards under supervision of the Department of Health and Human Services.

The PHS and NIH jointly initiated a study in Guatemala in 1946; it ran to 1948. They used prostitutes and, later direct inoculations, to infect prisoners, insane asylum patients, and soldiers, with syphilis. They tested penicillin on 1,500 infected people, 83 of whom died. The Surgeon General acknowledged that the study "... could not be conducted domestically." President Obama publicly apologized to the president of Guatemala in 2011. Marcia Angell, editor of the *New England Journal of Medicine* noted: "... researchers were still willing to conduct studies in developing countries that would not be acceptable in the U.S."

There is no doubt that patients, usually but not always suffering from schizophrenia, are more docile, as is a vegetable, following a frontal lobotomy (leucotomy). António Egas Moniz received the Noble Prize in medicine in 1949 for the development of this technique.

It is common and accepted practice to deceive volunteers or modestly paid subjects in psychology studies. In 1962, Milgram at Yale had an "experimenter," a volunteer "teacher," and a "learner," who was in on the research setup but pretended to also be an unsuspecting volunteer. The teachers thought they were participating in a study on memory and learning, when in fact it was their own obedience and respect for authority that was being tested.

The list goes on. The general community knew of these transgressions. Punishments or condemnations were infrequent. Such experiments were accepted as the unfortunate cost of doing science. The burden of proof rests with anyone arguing that these physicians derived sadistic pleasure from these experiments.

The Nazi regime in Germany conducted a series of medical experiments initially on their own citizens and subsequently on prisoners of war. Initially these were motivated by concerns of racial purity, subsequently by situations encountered in war.

The Law for the Prevention of Genetically Defective Progeny (1933) targeted weak-mindedness, schizophrenia, alcohol abuse, insanity, blindness, deafness, and physical deformities. From 1935 to 1939 about 300,000 German youths had been sterilized. Techniques at Auschwitz and Ravensbrück included surgery, drugs, and a secret X-ray zap from the seat of the chair while the prisoner was sitting completing forms.

At Dachau and Auschwitz, Ernst Holzlöhner and Sigmund Rascher with the Luftwaffe (1941) immersed subjects in tanks of ice cold water, recorded the water temperature, the time of death, and the temperature of the body. They tried various methods to revive the victims. Their results were presented at a *Medical Problems Arising from Sea and Winter* in 1942 to about a hundred attendees, most of whom were physicians. At Ravensbrück, the Wehrmacht (1942–1943) tried transplantations of bone, muscles, and nerves. No anesthesia was used. At Baranowicze in 1942, a Nazi Sicherheitsdienst (security officer) strapped a young boy to a chair and applied a mechanized hammer that came down upon his head every few seconds. The boy was driven insane.

At Auschwitz from 1943–1944, Josef Mengele oversaw experiments on 1,500 sets of twins. Dyes were injected into their eyes to see whether the change in color (presumably blue) was permanent. Others were sewn together in an attempt to create conjoined twins.

At Dachau (1942–1945), about a thousand prisoners were injected or bitten by mosquitoes. Various drugs were tested; half died. At Neuengamme, children were infected with *Mycobacterium tuberculosis*; axillary lymph nodes were removed for examination of the progression of tuberculosis.

At Sachsenhausen, exposure to mustard gas, bis(2-chloroethyl) sulfide, created severe burns; various treatments were tested. At Ravensbrück (1942–1943), prisoners' wounds were infected with defined strains of bacteria; blood flow was restricted, and various sulfonamide drugs were tested. At Dachau (1944), 90 Roma were given no food or water; only seawater was available. Many died. At Buchenwald (1943–1944), various poisons were secretly added to prisoners' food. They soon died or were killed and were autopsied. Phosphorous material from bombs was applied topically. At Dachau (1942), Sigmund Rascher subjected 200 subjects to the simulated pressure and temperature of 20,000 m altitude. Eighty died; the others were executed for autopsy.

The banality of evil — in August 1947, the captured German doctors were tried in "USA vs. Karl Brandt *et al.*" They offered as defense the precedent of Tuskegee studies and that there was no international law regarding medical experimentation. The Imperial Japanese Army's Unit 731 conducted similar studies. The results were kept classified; most of doctors were pardoned.

There were already significant documents defining human rights and freedoms:

1215 Magna Carta, England
1679 Habeas Corpus Act, England
1776 Declaration of Rights, Virginia
1789 Bill of Rights, U.S.
1789 Declaration of the Rights of Man and of the Citizen, France
1941 President Roosevelt (State of the Union), Four Freedoms — of speech, of religion, from fear, from want

These trials gave rise to the Nuremberg Code for human experimentation:

1) voluntary, informed consent
2) fruitful results of the experiment anticipated
3) based on previous animal experimentation
4) avoid unnecessary physical and mental suffering
5) no expectation that death or disabling injury will occur
6) degree of risk not exceed humanitarian importance
7) proper preparations and adequate facilities
8) conducted by qualified persons
9) subject should be able to terminate the experiment
10) scientist must be prepared to terminate the experiment

These examples have been presented neither to generate revulsion nor rectitude. They reflect outstanding, complex ethical questions that have profound implications:

Should prison inmates be paid or granted privileges for participation in studies deemed unacceptable for the general population?

Should children or adults be inoculated, at small but real risk, with the main beneficiary being the general population?

Should infants or young boys be relieved of their foreskins without their consent and should that tissue be subsequently used as a great source of fibroblasts?

Should tissue obtained from legitimate surgical procedures be used for subsequent experiments?

Should out-of-date blood, donated with the understanding that it be transfused to a fellow human in need, be used as a source of proteins for research?

Should probable risks and benefits of new drugs or procedures be weighted differently in industrial vs. developing nations?

Should information gathered for one legitimate medical or behavioral study be used by other investigators for other studies, even if personal identifiers are removed?

Should DNA samples taken from criminal suspects, later exonerated, be retained and accessed?

Should participants in psychology studies be deceived as to the true goal of the study?

Should "bio-markers" — fingerprints, iris scans, DNA sequences — be available to other scientists?

Informed consent is usually discussed in the context of medical experiments. However, the concept should be extended to consider the use of information about a person or a group. Examples include:

- Telephone taps with judicial approvable
- Ultra-sensitive listening device from public space
- Photography of a house from public space
- Photography (or listening) from mini-drones hovering a few feet overhead, soon available for $1,000
- Airport searches, pre-boarding
- Camera monitors in stores
- Camera monitors on public streets and highways
- Contents of web "searches" to tailor advertisements
- Addresses of e-mails and searches to establish patterns
- Content of e-mails with judicial approval
- Angela Merkel's shopping list
- Acquisition of biomarkers — finger prints, iris scans, DNA

What is a "reasonable expectation of privacy" and "informed consent" given rapidly evolving technologies?

Science, *per se*, cannot answer these questions. Scientists do not have special insight into ethical issues. However, all of these questions involve significant components of science. It seems reasonable that a more appropriate policy would be forthcoming if folks understood the relevant science and if they could rely on that science to be "true."

D3

Faith, Intelligent Design

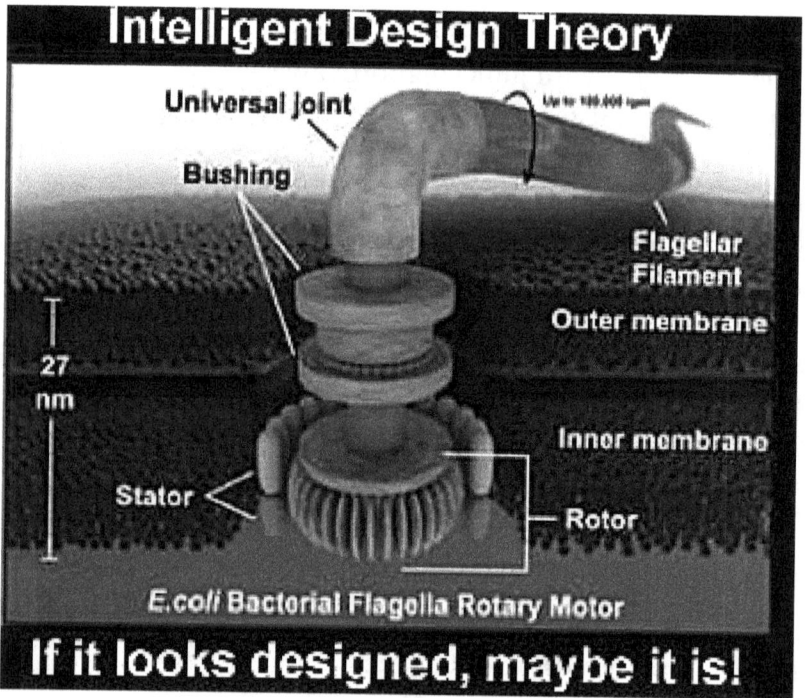

At great simplification, most religions have served three functions. Adherence of the community to beliefs and rituals can provide coherence and security; it might also discourage curiosity and innovation. Standards of behavior and ethics can be codified and more easily enforced; once engraved they are difficult to modify. Science has little to say to these two.

However, most religions also have creation stories and explanations for many natural phenomena. The Christian church(es) felt deeply threatened when its stories were challenged by heliocentrism and evolution but not by electricity and magnetism. For most of the faithful these concerns have been resolved; the attitudes of the church have changed markedly over two millennia. Today the Pontifical Academy of Sciences at the Vatican, and the leadership of most protestant denominations, are well informed and encourage scientific research for both its practical benefits and to better understand the workings of God.

Some might argue that one cannot be both committed to science and to a faith. Most rejoinders fall into two broad categories. The big bang was divinely inspired and energized; basic particles and forces were created. Then for the last sixteen billion years things have run their course without divine tweaking. One has faith that the big bang occurred and that certain rules, e.g. $E = m \cdot c^2$, obtain. The game plays out without divine intervention. Some scientists of faith simply do not question whether religion affects their science or whether science impacts their faith; the two are orthogonal.

Few question the unique status of humans and the sanctity of life. Most endorse the idea that women should be able to decide whether and when to bear children; this is not equivalent to endorsing a specific form of birth control. Most believe that society should protect children from the time of their birth; infanticide is no longer acceptable. At what point and in what way does the state intervene — ovulation, ejaculation, fertilization, implantation, fetal heart beat, viability at 28 weeks *ex utero*? Aristotle (384-322) speculated as to when the soul enters the fetus (Chapter B5). Do the scriptures of any faith explicitly address these issues; do people of faith have privileged insights?

There are conservative Christians, and people of other faiths, who champion a literal interpretation of scripture and some form of intelligent design. Most scientists are reluctant to engage these folks; that would just provide them voice and legitimacy. Yet, these believers influence education and medical policy.

One marvels at the diversity and complexity of life on earth. It seems inconceivable that these complex structures, behaviors, and molecules could have arisen by a mindless and soulless natural selection (Chapter C17). As Reverend Paley observed in 1836, if one found a watch by the road, would it be more reasonable to infer that it had been created through a "natural" process or by the hand of God? Many reasonable and critical people, including distinguished biologists, acknowledge that the living world has evolved; that is, it is different today from what it was in ages past. But surely there was some guiding influence? "Intelligent design" is now relegated to the backwaters of anti-intellectualism and primitive religiosity. However, Paley's argument and its refutation merit a more critical and sympathetic analysis, to understand the history of evolution by natural selection as well as a significant aspect of contemporary politics. Often the commitment to a political stance has less to do with critical evaluation of its logic and more to do with its symbolism and association with its adherents or opponents.

This view of creation and of intelligent design was generally accepted until the publication of *Origin of Species* in 1859. "I cannot possibly believe," wrote Darwin in 1868, "that a false theory would explain so many classes of facts." (Chapter C14). He acknowledged, "If it could be demonstrated that any complex organ existed which could not possibly have been formed by numerous, successive, slight modifications, my theory would absolutely break down" (a century before Popper, Chapter A10).

Richard Milner and Vittorio Maestro, senior editors of *Natural History* (2002) invited several contributors to address various aspects of intelligent design. Michael J. Behe, professor of Biochemistry at Lehigh University, described some molecular systems in the cell that require multiple components in order to function and called them "irreducibly complex." "The function of the system only appears when the system is essentially complete." "I have proposed that a better explanation is that such systems were deliberately designed by an intelligent agent." "An illustration of the concept of irreducible complexity is the mousetrap ... which needs all its parts to work." "It consists of 1) a flat wooden platform or base; 2) a metal hammer, which crushes the mouse; 3) a spring with extended ends to power the hammer; 4) a catch that releases the spring; and 5) a metal bar that connects to the catch and holds the hammer back. You can't catch a mouse with just a platform, then add a spring and catch a few more mice, then add a holding bar and catch a few more. All the pieces have to be in place before you catch any mice."

Kenneth Miller in the same volume responded with a description of "exaption"; that is, components of the system performed other functions in their early evolution. He dissected Behe's mouse trap:

"Take away the catch and the metal bar, and you may not have a mousetrap but you do have a three-part machine that makes a fully functional paper clip. Take away the spring, and you have a two-part key chain. The catch of some mousetraps could be used as a fishhook, and the wooden base as a paperweight."

On to the real world; the bacterial flagellum is often cited as an irreducibly complex system. The flagellum of *Salmonella typhimurium* consists of a:

1) "rotor-stator" that has several proteins and rotates the entire flagellum in response to an ion gradient.
2) type-III secretion system with nine core flagellar proteins
3) ATP-hydrolase

All together the flagellum contains about twenty different, genetically encoded proteins, all of which belong in one of two homolog families. There are many examples of much simpler systems based on only a few of these proteins, individual components of the mousetrap. As Darwin opined "... nature is prodigal in variety, though niggard in innovation."

Judge John E. Jones, in the Kitzmiller v. Dover (PA, board of education) trial, ruled that it is "... abundantly clear that the (school) Board's Intelligent Design Policy violates the Establishment Clause. In making this determination, we have addressed the seminal question of whether ID is science. We have concluded that it is not, and moreover that ID cannot uncouple itself from its creationist, and thus religious, antecedents." — 2006 http://www2.ncseweb.org/wp/?page_id=5

Governor Bobby Jindal, La, signed the "Academic Freedom Act" in 2008, which passed 94–3 in the state house; 35–0 in the state senate, "Some want only to teach intelligent design. Some want only to teach evolution. I think both views are wrong. As a parent, I want [children] to be presented with the best thinking. I don't want any facts or theories or explanations to be withheld from them because of political correctness. The way we are going to have smart and intelligent kids is exposing them to the very best science."

Recent standards for Texas schools incorporate talking points from the intelligent design literature, including doubt that the fossil record provides convincing evidence of evolution. "I think the new standards are wonderful," said Don McLeroy, chair of the Texas Board of Education who claimed that "dogmatism about evolution ... has sapped ... America's scientific soul." He argued that biology texts, to meet the new standards, should include "an evaluation of the sudden appearance of fossils" and "an explanation of stasis or how certain organisms stay the same over time." He also wanted the textbooks to declare there is no "scientific explanation for the origin of life" and that "unguided natural processes cannot account for the complexity of the cell." *Science* (2009) **324**, 25.

The Center for Science and Culture calls its strategy the "Wedge," because it wants to liberate science from "atheistic naturalism." CSC has an aggressive public relations program, which includes conferences that they or their supporters organize; popular books and articles; recruitment of students through university lectures sponsored by campus ministries; and cultivation of alliances with conservative Christians and influential political figures.

Michael Reiss is a biologist and an ordained minister; he was the Royal Society's Director of Education from 2006–2008. In a public lecture at a national science festival, 10 September 2008, he said that creationism should not be covered in the science curriculum, "I feel that creationism is best seen by science teachers not as a misconception but as a world view." Several members of the Royal Society called for his dismissal, "Some of Professor Michael Reiss's recent comments, on the issue of creationism in schools, while speaking as the Royal Society's director of education, were open to misinterpretation." He resigned 16 September 2008. This game is not for sissies.

Joseph Ratzinger, Pope Benedict XVI, gave an address, "*Faith, Reason and the University — Memories and Reflections*" at the University of Regensburg, 12 September 2006, in which he called for dialog between different religions and between religion and science. "This gives rise to two principles which are crucial for the issue we have raised. First, only the kind of certainty resulting from the interplay of mathematical and empirical elements can be considered scientific. Anything that would claim to be science must be measured against this criterion. Hence the human sciences, such as history, psychology, sociology, and philosophy, attempt to conform themselves to this canon of scientificity. A second point, which is important for our reflections, is that by its very nature this method excludes the question of God, making it appear an unscientific or pre-scientific question. Consequently, we are faced with a reduction of the radius of science and reason, one which needs to be questioned." "Attempts to construct an ethic from the rules of evolution or from psychology and sociology end up being simply inadequate." "Modern scientific reason quite simply has to accept the rational structure of matter and the correspondence between our spirit and the prevailing rational structures of nature as a given, on which its methodology has to be based. Yet the question why this has to be so is a real question, and one which has to be remanded by the natural sciences to other modes and planes of thought: to philosophy and theology."

D4

Art

Lascaux
Dordogne
(cave painting
~17,300 ybp).

Art and Science seem inherently different, yet one might argue that the urge to do art and the urge to do science (pure and applied) derive from the same basic intellectual processes. To what extent and in what ways did and do art and science interact?

Art portrays many themes — sculptures of DNA, photographs of Mars — taken from science. Art uses many of the devices — new pigments, digitalization of images — provided by science. The more interesting question is how art has informed science?

Many scientists describe their work in the language of aesthetics — an elegant derivation, delightful symmetry, lovely pattern. For those in the field this seems natural and appropriate. Although a sense of aesthetics is essential, not only for appreciating art, but for doing art, there is also the execution, the interface between the aesthetic and the mechanical.

Understanding ratios of frequencies of vibrating strings contributed to the development of early number theory. Neuroscientists are trying to understand the appeal of the octave or of the fifth in terms of which hair cells in the cochlea respond. One might consider a parallel analysis of the appeal of primary colors and the absorption maxima of the three rhodopsins of cone cells.

The golden mean gave a neat equation and generated an early example of an irrational number: $(a + b)/a = a/b = 1.61803399\ldots$ ($\sim 8/5$). Science fiction has predicted many things, well beyond the understanding or capabilities of its contemporary sciences. A few of these predictions were prescient; to what end?

Perhaps these examples miss the point. Art, in its myriad forms, may inspire the scientist at a deeper, and more difficult to analyze level.

All of the seven, nominal art forms — painting, sculpture, architecture, literature, theater, music, and dance — have used materials or themes derived from science. The more challenging question is: How has art informed science?

Most people had and have a sense of esthetics. People recognize beauty, as beautiful, in many cultures.

"The soul is awestricken and shudders at the sight of the beautiful, for it feels that something is evoked in it that was not imparted to it from without by the senses, but has always been already laid down there in the deeply unconscious region." — Plato, *Phaedrus*. "Beauty is the proper conformity of the parts to one another and to the whole." — Francis Bacon. "Beauty in things exists in the mind which contemplates them." — David Hume.

People recognize art, as art, in many cultures. Certainly humans had and have an inherent urge to do art — rock art of South Africa and Lascaux, Venus of Willendorf, temples, the *Iliad*, tribal songs and dances of pre-agricultural people. The urge to do science seems universal — math, astronomy and astrology, complex structures and weapons, botany and healing. Are doing art and doing science reflections of the same curiosity and creativity? Would the same regions of the brain light up when on task?

Certainly scientists find beauty in science. They speak of a lovely result or an elegant derivation or a beautiful electron micrograph; however, one can hardly argue that this beauty is essential to the practice. Does the beauty of nature affect discovery, analysis, and understanding? Man would have looked for patterns in astronomy or anatomy anyway; is there beauty or art in an ephemeris or a ripening cadaver? To what extent did esthetics inspire the formulations of Newton, Darwin, or Maxwell? Chandrasekhar, (Nobel Laureate, 1983) explored *Truth and Beauty: Aesthetics and Motivations in Science* (1990).

Beauty is truth, truth beauty — that is all
Ye know on earth, and all ye need to know.

— Keats

Adrian Forty in *Words and Buildings: A Vocabulary of Modern Architecture* (2000) discussed the range of influences on architecture: "... since the scientific metaphors employed in architecture are drawn from such a diversity of scientific fields, from natural sciences as well as physical sciences and mathematics, the cumulative effect is to suggest the unlikeness of architecture from science in general." John Ruskin observed: "... art which so disposes and adorns the edifices raised by men ... that the sight of them ..." contributes "... to his mental health, power, and pleasure." Vitruvius, the Roman architect, stressed three criteria: Durability — it should stand up robustly and remain in good condition. Utility — it should be useful and function well for the people using it. Beauty — it should delight people and raise their spirits. Alberti elaborated rules of proportion that govern the idealized human figure and said that beauty was an inherent part of an object and was based on universal, recognizable truths. For example, he championed the golden mean: $(a + b)/a = a/b = 1.61803399... (\sim 8/5)$, one example of art having inspired mathematics.

Jonah Lehrer in *Proust Was a Neuroscientist* (2007) argued that: "This book is about artists who anticipated the discoveries of neuroscientists." (See accusation of "self-plagiarism," Chapter D1). Of several examples Lehrer noted that only four tastes were traditionally recognized in the West, as described by Democritus:

sweet   "round and large in their atoms"
sour   "large in its atoms but rough, angular and not spherical"
salt   "by isosceles atoms"
bitter   "spherical, smooth, scalene and small"

However, Auguste Escoffier (chef at César Ritz) in his *Guide culinaire* (1903) described the unique taste of meat, now attributed to the amino acid, L-glutamate, as the fifth. Umami was long recognized in Japan. Five different taste receptors have now been identified on the tongue.

Yves Christen, a neuroscientist, explored the *Biologie de l'esthétique*, (Actualités de l'académie des Beaux-Arts, Paris). He described bird songs and plumage and asked whether a sense of beauty exists in nature? He asked "why beauty?" Does the plumage of male peacocks provide a selective advantage for the individual but a burden for the species? Does the hen, or any female subject to courtship, regard male antics as beautiful? Fernando Nottebohm identified specific neurons that are involved in learning bird songs; chicks can be taught to sing different tunes. Might one develop a neuro-psychology of beauty?

Pythagoras ($\sim$576–$\sim$495 B.C.) argued that "... number is the ruler of forms and ideas and the cause of gods and demons." He gave the first proof $a^2+b^2 = c^2$ for a right triangle and realized that $2^{0.5}$ is irrational (Chapter B2). His analysis of vibrating strings and musical intervals is certainly one example of music having inspired mathematics. The music of the heavens should be expressed as ratio of integers: 2:1, 3:2, 4:3, etc.

In contemporary notation A2 is 110 hertz (vibrations per second). In an equal temper scale, normalized to 100 for ease of illustration:

A(2)	100.0			unison	
A#	106.0			minor second (half note)	
B	112.3	9/8 =	112.5	major second (whole note)	
C	119.0	6/5 =	120.0	minor third	
C#	126.0	5/4 =	125.0	major third	$(9/8)^2 = 126.63$
D	133.5	4/3 =	133.3	perfect fourth	
D#	141.4	45/32	140.6	augmented fourth	$(9/8)^3 = 142.38$ (or diminished fifth)
E	149.9	3/2 =	150.0	perfect fifth	
F	158.8	8/5 =	160.0	minor sixth	$(9/8)^4 = 160.18$

F#	168.2	27/16	168.7	major sixth	
G	178.2	9/5 =	180.0	minor seventh	$(9/8)^5 = 180.20$
G#	188.8	15/8	187.9	major seventh	
A(3)	200.0	2/1 =	200.0	octave	$(9/8)^6 = 202.73$

A whole interval (W) is 112.3 (~9/8, 112.5); a half interval (H) is 106.0 $(9/8)^{0.5}$.

Western modes consist of five <u>W</u>hole notes (intervals) and two <u>H</u>alf notes.

The "natural scale" consists of W H W W W H W intervals in an octave. The "melodic" consists of: W W H W W W H going up and W H W W W H W going down to lower frequency. The "harmonic" consists of: W H X H W X H in which "X" is halfway between a whole and a half, i.e. $(9/8)^{0.75}$.

The modes of the natural scale are:

Ionian	W-W-H-W-W-W-H
Dorian	W-H-W-W-W-H-W
Phrygian	H-W-W-W-H-W-W
Lydian	W-W-W-H-W-W-H
Mixolydian	W-W-H-W-W-H-W
Aeolian	W-H-W-W-H-W-W
Locrian	H-W-W-H-W-W-W (intervals)

This was hardly the formulation of Pythagoras, but he did get the ball rolling, or more appropriately, the strings vibrating. Why five whole and two half intervals? Or six wholes? Or four wholes and four halves? Why this sequence of the five wholes and the two halves in the natural scale? Chinese or African music can only be approximated by these scales. One might wonder why the West locked in on these conventions? Can one deduce anything about their society or physiology?

The great art critic Bernard Berenson wrote in 1896:

> Leonardo (da Vinci) is the one artist of whom it may be said with perfect literalness: Nothing that he touched but turned into a thing of eternal beauty. Whether it be the cross section of a skull, the structure of a weed, or a study of muscles, he, with

his feeling for line and for light and shade, forever transmuted
it into life-communicating values.

Da Vinci described the four cavities of the heart, argued
that ventricular valves prevent regurgitation, and demonstrated
that under pressure no air is transmitted to the left auricle of the
heart from bronchi (Chapter C3). A few of his fantasies included:
the helicopter, hang glider, tank, solar power, calculator, double-
hulled vessel, hydraulic pump, finned mortar shell, and steam cannon. How did his art inspire his science?

"Science fiction" and "fantasy" are closely related but the distinction — in fable, book, or film — is valuable: Science fiction
describes the same world with new gadgets or anomalies, e.g. men
invent devices that permit flight.

Frankenstein (1818)	Mary Shelley
Voyage to the Center	
of the Earth (1871)	Jules Verne
The Time Machine (1895)	H.G. Wells
I, Robot (1950)	Isaac Asimov
A Sound of Thunder (1953)	Ray Bradbury
Stranger in a Strange Land (1961)	Robert Heinlein
Star Trek (1966)	Gene Roddenberry
2001: A Space Odyssey (1968)	Arthur C. Clarke
	& Stanley Kubrick
Slaughterhouse-Five (1969)	Kurt Vonnegut
Bladerunner (1982)	Ridley Scott
Neuromancer (1984)	William Gibson

"Fantasy" describes new worlds with different "laws," e.g. men
grow wings.

The Odyssey	
Old Testament	
One Thousand and One Nights	
New Atlantis (1623)	Francis Bacon

The Tempest (1623)	William Shakespeare
Gulliver's Travels (1726)	Jonathan Swift
Alice's Adventures in Wonderland (1865)	Lewis Carroll
A Connecticut Yankee in King Arthur's Court (1889)	Mark Twain
Walden Two (1948)	B. F. Skinner
The Lord of the Rings (1954)	J. R. R. Tolkien
One Hundred Years of Solitude (1967)	Gabriel García Márquez
Harry Potter and the Philosopher's Stone (1997)	J.K. Rowling

One might quibble about this division and the choice of works listed. No doubt these stories inspired both adults and youngsters, some of whom became scientists. However, the question relevant to this discussion is whether, in any more specific sense, this art advanced science?

D5

Global Warming

Careser glacier
August/1933

Careser glacier
August/2012

Global warming is one of many examples of a challenge to society that has science, both basic and applied, at its core. How to think about it? First one measures, or infers, global and local temperatures over the past thousand, million, or billion years. One measures immediate indicators of temperature — glaciers, ice packs, ocean levels — as well as secondary correlates — vegetation, etc. These measurements may be technically challenging; however, they are subject to verification and refinement.

One then seeks correlates, such as fluctuations in solar output or massive volcanic activity or bolide collisions. Arrhenius in 1896, after an earlier suggestion by Fourier, proposed that the Earth's mean temperature should increase by $\sim 4°$ for each doubling of the concentration of CO_2 in the atmosphere. There may be causes without precedent, such as human activities like burning fossil fuel. Even people without a political agenda might debate the causal relationships of these many variables. Those with financial or ideological commitments are seldom so objective.

One aspect of the policy debate is whether to focus resources — by emotion, incentive, law, or treaty — on reducing emission of CO_2 from fossil fuels. Or, one can accept projected levels of CO_2 emission and attendant changes in climate and invest resources in dealing with predicted, altered patterns of agriculture, navigation, construction, etc. The relevance of such economic and political analyses depends critically on these projections and their underlying models.

How should one analyze global warming? What questions should be asked? How should the various desiderata be weighted? To what extent are the various opinions on global warming subjective and to what extent correlated with other concerns, such as health care and conservation? As noted in other contexts, science does not provide an explicit answer to these questions; however, a wise policy should recognize solid science.

In the same general period that scientists first investigated ice ages and climate change, Joseph Fourier (1768–1830), in 1824, realized that the Earth's atmosphere keeps the planet warmer than would be the case if it had no atmosphere. He made the first calculations of this warming effect. Fourier recognized that the atmosphere transmitted visible light from the sun to the Earth's surface. The Earth then absorbed some of this light energy and emitted infrared radiation in response. However, the atmosphere absorbs much of this infrared, thereby increasing the fraction of solar energy that is absorbed. He also suspected that human activities could influence climate, although he focused primarily on land use changes. In 1827, Fourier stated:

> The establishment and progress of human societies, the action of natural forces, can notably change, and in vast regions, the state of the surface, the distribution of water and the great movements of the air. Such effects are able to make to vary, in the course of many centuries, the average degree of heat; because the analytic expressions contain coefficients relating to the state of the surface and which greatly influence the temperature.

John Tyndall took Fourier's work one step further in 1864, when he investigated the absorption of infrared radiation in different gases. He found that water vapor, hydrocarbons like methane (CH_4), and carbon dioxide (CO_2) strongly absorb infrared radiation of the wavelengths emitted by the Earth.

Svante Arrhenius (1859–1927) used Langley's observations of increased infrared absorption of moon rays passing through the atmosphere at a low angle to estimate an atmospheric cooling effect from a future decrease of CO_2. He realized that the cooler atmosphere would hold less water vapor (another greenhouse gas) and calculated the additional cooling effect. He also realized the cooling would increase snow and ice cover at high latitudes, making the planet reflect more sunlight and thus cooling further, as James Croll had hypothesized. Overall Arrhenius calculated that cutting CO_2 in half would suffice to produce an ice age, not an unreasonable concern for someone living in Sweden. He further calculated that a doubling of atmospheric CO_2 would give a total global warming of 5–6°C.

Viewed a century later, Arrhenius got it right. Appropriately, scientists are now concerned with details such as spatial and temporal fluctuations in temperature over the globe. Historical patterns of ice ages and warm periods are, sort of, correlated with Milankovitch cycles that describe the tilt of the Earth's axis and its precession as well as the ellipticity of its orbit and its precession (Chapter B3). The thermometers — polar ice packs and glaciation — that integrate global temperatures over decades and centuries leave no doubt that the Earth has been warming over the past century; ask any mountaineer.

It is quite appropriate that countries reduce their emissions of CO_2 as driven by incentive, fiat, and/or treaty. However global warming is already occurring and will continue to occur. Society is well advised to anticipate its effects and prepare as best it can. There will be losers and winners; however, one can safely anticipate:

Sea levels will rise over a meter during the 21st century and weather fluctuations (storms) will increase. Low lying areas, within a few meters of present sea level, should be evacuated and/or hardened well before, not after, they are inundated; new construction should be restricted.

Some areas, for instance the Sahel, will experience desertification. Their agricultures will no longer be able to feed their people. These nations should be exploring new crops better suited to these (partially) predictable changes.

Animals and plants in the past have adapted to many changes of climate over millennia. Changes over decades will quickly exhaust the variation within their existing gene pools (Chapter C11); it takes much longer to acquire and to select new mutations. Some animals can migrate; plants move more slowly.

There will be strong immigration pressures on more temperate and prosperous nations. Strong fences and detention centers can be only part of the answer. Some nations may experience longer growing seasons and greater agricultural productivity. These "winners" can anticipate demands to share their largesse. New, (almost) year round shipping lanes in the Artic will alter patterns of world trade.

Now is the time for futurologists to belly up.

D6

Free Will

Free will is one of those terms frequently used but difficult to define. This makes it no less important in analysis nor less valuable in execution. Free will is often contrasted with determinism, behavior guided by an algorithm, explicit or implicit.

A great deal of research in both animal behavior and human psychology is devoted to deciphering these algorithms — their details, development, and effectiveness. Viewed in this perspective free will is noise, deviation from the norm.

Under some circumstances these deviations are destructive and reduce the efficiency or productivity of a system. In other situations, free will is essential to the creative process. From a biological perspective one might ask whether free will is adaptive.

An exploration of the evolution and ramifications of free will and of determinism might enable us to better understand the subtleties of both and their appropriate interactions.

Free will is often contrasted with determinism. This implies a binary division, one or the other. It is difficult to sort or categorize acts of free will; the investment of intellectual or emotional analysis runs on a continuum. Sometimes the consequences of these decisions have profound implications, not anticipated at the time the act of free will was made; most are trivial. One might anticipate the importance of the decision and assign more neurons to the big ones; this, of course, involves another decision. The basic question is whether all decisions can be evaluated within the same conceptual framework.

This question has challenged philosophers and prophets for millennia. As discussed in the context of two cultures (Chapter A12), it seems a stretch to argue that biology will provide answers to fundamental questions of aesthetics and ethics. However, it is reasonable to ask biology to flag fallacious arguments.

Some people believe that obese people or homosexuals are "born that way." Setting aside evolutionary arguments concerning adaptive behavior, others argue that behaving the "right" way reflects the exercise of free will.

Many decisions actually involve a cluster or sequence of component evaluations or decisions. The wolf has to "decide" which buffalo might offer the least resistance; she must decide how many of her pack will join, given a path through the brush; she must consider how far she has run from her hungry pups. Are these "preliminary evaluations" inherently different from the big decision of which buffalo to take down?

It must be exhausting to exercise free will when making each of thousands of decisions every day. Some reflect parasympathetic circuits not involving the brain, e.g. to retain urine while asleep. This can be over-ridden by a conscious decision, exercise of free will, after getting to the bathroom. Brushing one's teeth is pretty far down on the list of the day's intellectual challenges. Choosing attire is a bit more of a challenge, if this is for a job interview, of major import. The vast majority of these decisions seem to be of little significance; however, that left turn into oncoming traffic might have fatal consequences. The sum of trivial decisions may set a pattern of major import.

Some would define free will as a uniquely human attribute; this certainly simplifies subsequent discussions. However, many examples of the exercise of free will can be applied to animals. As noted, wolves decide which buffalo to cull from a herd and chimpanzees choose other chimps to include or exclude from the clan. In addition to a species dimension there is a temporal dimension. Neanderthals decided when and how to fashion simple axes. Further back, the first proto-bacterium to regulate gene expression made a decision. One might dismiss this as sophistry; yet the point is how to define or characterize a decision. This in turn raises the question of how a decision(s) is (are) made. The basic assumption is that the exercise of free will is evinced by the decisions made. Or can one monitor blood pressure or brain waves to detect the decision to consider a decision?

Can training of a dog be considered to involve the transfer of its decisions involving free will to the determinism of a learned behavior? If that prize-winning poodle were released to the wild, would it be at great disadvantage relative to a junkyard mutt? To what extent do we want to train our students to do well on algorithmic exams and to what extent to retain free will and creativity? Is it a zero-sum game?

Free will, as so characterized, is not a uniquely human attribute. Can the analysis of free will in humans be extended to other animals? If one discusses contemporary folks in western societies, should one extend the conclusions to other humans?

Organisms have evolved adaptive behavior. Bacteria divide and plants flower under certain conditions of light, temperature, nutrients, and moisture, etc. Bees take nectar back to their hives and mammals nurse their young. Averaged over individuals and time these actions assure the survival of the species. Few would argue that these decisions reflect the exercise of free will.

One is left to sympathize with the frustrated philosopher. Can free will be relegated to the autonomic nervous system by an act of free will, perhaps unbeknownst to the willer? Can acts of free will be ranked by anticipated or actual consequences? Is free will a uniquely human attribute? Can the acquisition of free will be correlated with the transition from the *Pan/Homo* ancestor to *Homo*?

However, this is not what is usually meant by free will. Should I kill one stranger to save two? Should I marry Amy or Beth? These heuristics capture our attention; however, are these big decisions inherently different from the thousands of seemingly trivial ones? Do societal norms, consciously or subconsciously, impact big free will any more than small free will? Do restrictions on one class of decision have a carry-over impact on others? Inherent to these examples is the implication that the decisions reflecting free will matter. Life with Amy will be different than life with Beth; my offspring will carry different genes.

Determinism involves the past experiences and the genes of the organism. And determinism considers the external circumstances that obtain at the time of the decision. If the organism is on auto-pilot, it might make a series of complex decisions without the exercise of free will. That is, it (its species) has evolved mechanisms to monitor various external and internal parameters and make decisions that are pretty good, most of the time. By definition free will monitors this ensemble of deterministic decisions, like a teacher watching a playground of youngsters. It intervenes only when some observable parameter scores a significant deviation. That tipping point might have evolved; it might be changed based on experience, all without free will.

In this model free will addresses only the biggies and, more important, it can change the definition of the big ones, rapidly and wisely without the exercise of free will. This definition allows one to extend the analysis to any entity at any time. Is this reasonable or fruitful?

Most people, knowingly or not, establish daily rituals. Certainly (some of) the justification for such rituals is to free one from the burden of trivial decision making. How to sort out the free will to be delegated from the free will to be exercised? This leads to a decision inside a decision, inside ... a Matryoshka doll.

Inherent tendencies toward certain behaviors are well documented in, for instance, the "Minnesota (Identical) Twin Family Study" (Joseph, 2004). Ever more behavioral and intellectual characteristics are being found to correlate with genotype, i.e. specific sequences of DNA. This implies that the tipping point(s) for the exercise of free will in decision making is the same in both twins. However, many of the decisions, e.g. mate and career selection, would involve free will. And these correlate between twins. Is it really free will if influenced by DNA? Some would reply "no problem" if heredity affects free will. Others consider this totally contradictory; by definition free will is beyond the reach of grubby purines and pyrimidines.

However, the situation is a bit more complex. Conditions *in utero*, whether (partially) determined by the genotype of the mother or by the environment, may significantly affect the future behavior of the fetus, or behaviors of the pair of fetuses. The societal responses to these observations vary enormously. As is often the case, one's view of the "evidence" is easily influenced by one's attitude concerning norms and policy.

Such discussions of genes and environment beg the question of what, if anything, gives us choice or gives us character. Is chance the third component? Is free will reduced to genes, environment, and a bit of luck?

The seeming conflict between determinism and free will has in varying contexts engaged philosophers for millennia. Theological determinism is the idea that God determines all that humans do, either by knowing their actions in advance, via some form of omniscience or by decreeing their actions in advance. The problem of free will, in this context, is the problem of how our actions can be free if there is a Being who has determined them for us in advance.

Psalm 16:4 — "Thou wilt shew me the path of life . . ." — implies that once one has committed, the subsequent decisions associated with free will are, to some extent, already made. This, of course, begs the question of the free will of commitment. One then asks: "Why should I strive and suffer to do good if my life course has been pre-ordained?" The answer takes various forms, among them: "This is God's way of testing your soul and determining your status in heaven." The contradictions in this argument are self-evident.

Other forms of determinism include: cultural determinism and psychological determinism. Combinations and syntheses of determinist theses, e.g. bio-environmental determinism, are even more common.

It is well established that both heredity and circumstances, including physical and psychological trauma, can affect behavior — both "programmed decisions" and free will. Legal scholars have wrestled for centuries with the concept of insanity, e.g. the U.S. "Model Penal Code," and how to determine whether a suspect should stand trial (Porter, 2002). If he does stand trial, it is assumed that he exercised free will — no gradations, yes or no. However, if convicted, extenuating circumstances may be considered during sentencing.

Thomas Carlyle emphasized the importance of great men; their acts of free will determine the course of history (*On Heroes, Hero-Worship and the Heroic in History*, 1888). In contrast Otto von Bismarck argued that leaders should recognize the tide of history, itself an act of free will, and ally with the prevailing winds.

Some people are disinclined to believe that behaviors in general and intelligence in particular are, at least in part, genetically determined, because the valid extrapolation is that one geographic or cultural subgroup of humans might be inherently more intelligent than another. Further, one might in the near future test or even manipulate a fetus to determine or to enhance its projected intelligence or other trait.

Arguments in philosophy are often presented as binary choices, black or white. The free will debate is better understood if specific decisions are viewed as weighted combination of chance, determinism, and free will.

Are we asking the right question?

Author Index

Subject Index

www.ingramcontent.com/pod-product-compliance
Lightning Source LLC
Chambersburg PA
CBHW061620220326
41598CB00026BA/3828